Contents

Introduction 1

1 **Introduction to Communications Switching** 3
   Objective 3
   Communications Traffic 4
   Circuit Switching and Message Switching 5
   Security 6
   Switching Terminals and Switching Links 6
   Switching in Space and Time 7
   Space Division Switching 8
   Time Division Switching 11
   Signalling, Routing and Control 13
   Chapter Summary 18
   Exercises 19

2 **Introduction to Communications Transmission** 20
   Wire Transmission 20
   Other Transmission Media 22
   Analog Signalling 22
   Transmission Quality 23
   In Perspective: the Relationship between Transmission and
      Switching 26
   Chapter Summary 28
   Exercises 28

3 **Pulse Code Modulation** 29
   Example PCM System 29
   Practical PCM Systems 37
   Line Coding 46
   Chapter Summary 48
   Exercises 48

4 **Error Sources and Prevention** 50
   Quantisation 50
   Practical Coding Laws 52
   Other PCM Coding Techniques 58
   Filtration 60
   Interference and Crosstalk 62
   Timing 64
   Chapter Summary 68
   Exercises 68

5 Digital Exchanges and Switching 70
Introduction 70
Switching: Analog Techniques and Blocking 71
Switching: the Digital Requirement 76
Time Switching 78
Time Switching in Practice 86
Chapter Summary 97
Exercises 97

6 Digital Exchanges and Control 99
Introduction 99
The Trend towards Centralisation 100
Control Security 106
Control Software 112
Overload Control 124
Chapter Summary 125
Exercises 126

7 The Digital Network 128
Network Synchronisation 128
Signalling 134
Digital Signalling 137
Common Channel Signalling Systems 145
Open Systems Interconnection—the ISO Reference Model 160
Chapter Summary 162
Exercises 163

8 Digital Frontiers 165
Line and Trunk Scanning 166
The Subscriber's Line Interface 166
The Trunk Network Interface 173
Line and Trunk Testing 175
Integrated Networks 175
The Exchange Building 176
Chapter Summary 180
Exercises 180

9 System Review and the Next Generation 182
The Shape of Systems 182
The Next Generation 186
Post-digital Communications Switching 192
Annex to Chapter 9: Survey of Digital Switching Systems 193

Answers to Exercises 199
Glossary 207
Index 210

Introduction to Digital Communications Switching

John Ronayne
BSc, CEng, FIEE
Communications Consultant

Pitman

PITMAN PUBLISHING LIMITED
128 Long Acre London WC2E 9AN

A Longman Group Company

© J P Ronayne 1986

First published in Great Britain 1986

All rights reserved. No part of this publication may be reproduced, stored in a retrieval system, or transmitted, in any form or by any means, electronic, mechanical, photocopying, recording and/or otherwise without the prior written permission of the publishers. This book may not be lent, resold, hired out or otherwise disposed of by way of trade in any form of binding or cover other than that in which it is published, without the prior consent of the publishers.

ISBN 0 273 02178 8

Dedication: to Mary

Acknowledgements To allot credit to all my many sources is difficult. The various references acknowledge my principal technical creditors. I have a greater debt to those who helped me to grow in this fascinating discipline. Among my many mentors I wish to name the following few:

L J (Jimmie) Brooks.
L R (Billie) Brown.
G C (Bottle) Hartley.
G A Conway.
D A Weir.
S M Schreiner.
R H Hayward.
S G W (Johnny) Johnstone.

JR 1986

Printed and bound in Great Britain at The Bath Press, Avon

Introduction

By the time that the telephone was invented in 1875 there was already a considerable body of theoretical and empirical knowledge on the problems of electrical transmission of information. Up to that time the only information transmitted electrically was digital information sent over what was, even then, a world-wide network of telegraph circuits. Codes, such as Morse Code, were devised to improve the quality of the received information despite the losses in quality introduced by the electrical transmission circuit. The introduction of voice transmission presented a completely new set of problems since now the absolute quality of the analog signal was of interest and not just a quality sufficient to detect the presence or absence of a signal.

Within a very few years of the invention of the telephone, the concept of switching was introduced and employed, at first manually but very soon using automatic exchanges. Immediately there was a divergence of the two arts of switching and transmission; the one largely empirical, the other resting on an increasing and sophisticated body of scientific and mathematical knowledge.

Speech transmission over long distances was made possible first by the loading coil and distances were increased by the introduction of thermionic valves. The thermionic valve also made possible high-speed switching of voice and data (telegraph) signals and, by the 1930s, widespread use was being made of multiplexed transmission methods where many voice and data connections could be carried over the same transmission circuit. Methods of multiplexing used then, and still in use to-day, included frequency division multiplexing (FDM), by amplitude-modulation or phase-modulation of the carrier frequency. Time division multiplexing (TDM) was considered but not at that time used for voice because of the difficulty of devising electronic switches which had sufficient difference between their switched-on and switched-off states.

The art of switching proceeded along parallel but independent lines. The earlier switching machines utilised distributed control in a step-by-step fashion, the connection being built up stage by stage as the caller dialled successive digits of the telephone number. The telephone number, therefore, exactly expressed the switching route between caller and correspondent. As automatic switching was introduced to larger conurbations, this became unsatisfactory and methods of time-divided common control were introduced.

By the late 1930s switching engineers and transmission experts were uncomfortably aware that both of them were engaged in switching traffic; in space division in the telephone exchange and in frequency division at each end of the multiplexed transmission link. Neither group of experts liked this unnecessary duplication but could see no solution.

A new form of transmission medium was introduced in the 1930s, that of microwave radio. Characteristics of this medium are a large available bandwidth but severe problems of noise and distortion. The investigation of this problem by a team of engineers at Laboritaire Central Telephonique in Paris resulted in the filing of a patent by Alec Reeves in 1938 suggesting a method of Pulse Code Modulation (PCM) which, by converting analog signals into digital code, might overcome the problems of noise and distortion. The electronic devices did not then exist to introduce PCM commercially and the idea lay dormant until the invention of the transistor in 1948 made it possible. Shannon's information theory expositions, also in 1948, made the potential of PCM for both transmission and switching apparent.

The combination of PCM and time division multiplex, therefore, had at last provided the tool to bring the arts of switching and transmission together. It allows the switching inherent in both arts (in switching, line and trunk to line and trunk; in transmission, circuit connection to transmission channel) to be performed once and once only for both purposes.

The development since 1948 has been rapid in the transmission area and somewhat slower in the switching field. Transmission engineers have been dealing with concentrated traffic and have been able to show immediate savings. Early PCM transmission systems were designed as direct replacements of FDM systems and employed a modularity similar to the systems they replaced. The US standard remains 24 channel to this day. The later advent of PCM switching has imposed the more universal binary modularity and the system favoured by World Telecommunications is now based on 32 channels. The coding standards generally used world-wide require a channel capacity of 64 kbits/sec to carry the standard telephony voice range of 300–3400 Hz.

To introduce PCM switching to the telephone exchange, beyond the large trunk exchanges switching concentrated traffic, the switching engineers have had to overcome the problems of providing PCM techniques in a form cheap enough to be used in the concentration portions of the exchange. They have been assisted in this by the advent of large-scale integration (LSI) and by the foreseen demand by subscribers for more sophisticated services, making the provision of a 64 kbits/sec circuit to the subscriber an economical proposition.

This is not the end of the story but certainly we have passed a major milestone in the continuing story of communication. Transmission and switching, divorced for 120 years, have merged and present to the user almost unlimited capability for world-wide communication of speech and data. It is this which makes possible the information society whose birth we are witnessing.

1 Introduction to Communications Switching

To appreciate the role and the techniques of digital switching it is necessary first to outline the purpose and objective of communications switching. The transition from analog space division switching to digital time division switching is a change in technology which does not alter the architecture and concepts inherent in communications switching. These concepts must be appreciated before discussing the technology of digital switching.

Objective

The objective of the communications switch is to connect any calling inlet to any wanted free outlet at any time. The general method by means of which this function is performed can be explained by referring to *fig. 1.1* and by considering the functions performed by a telephone operator sitting at an old-fashioned manual switchboard (*fig. 1.2*).

On the manual switchboard, lines and trunks are terminated on jack sockets with associated calling indicators. The terminations are multipled over a number of positions so that an individual trunk might appear, typically, on every third position. An individual line, having a much lower calling rate, would appear again on every sixth or ninth position. Operators are allowed to stretch over to plug into a trunk on a neighbouring position. The operator continuously scans the board for the appearance of new calls and plugs into the calling line with a free cord circuit, throwing the cord circuit speech key to speak to the caller. The operator receives the caller's verbal instructions,

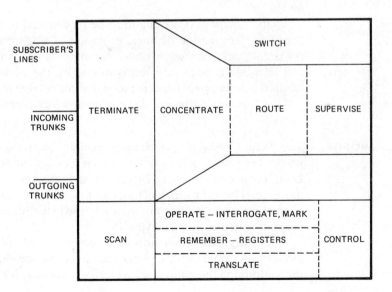

Fig. 1.1 Functions of a communications switch

4 Introduction to Digital Communications Switching

Fig. 1.2 Operators working at an auto-manual switchboard

translates them, using tables and directories where necessary, and sets up the wanted connection by plugging the other end of the cord circuit into the wanted line or trunk and passing on instructions where necessary to intermediate operators. Finally, ringing is applied to the called line and, on answer, the operator retires from the connection but monitors the indicators for a release signal in order to unplug the connection and make out a charge ticket for the call.

All the functions described must be reproduced in the automatic exchange (*fig. 1.1*): termination of lines, provision of ringing and electrical power to the telephones, scanning for calls, reception of instructions, memorising them and translating them into terms usable by the network. Lastly the switch fulfills the functions of concentration and connection provided by the multiplex and the operator's arms moving cord circuit jack plugs.

Communications Traffic

The picture of *fig. 1.2* gives an immediate impression of what we mean by traffic: very many calls from diverse sources and going to diverse destinations being connected at once. Moreover, subscribers are free to make calls at any time. The switching machines and transmission facilities must be on duty all the time and must be dimensioned to switch both expected and unexpected peaks of traffic economically.

Traffic originates from individual subscribers where calling rates are generally low. A suburban telephone exchange will be dimensioned typically on the assumption that individual subscribers are busy for about 4% of the busiest hour of the day. By contrast, the concentration switches of the exchanges ensure that trunks are busy for from 60% to 90% of the busiest hour. The amount of traffic that can be loaded onto the trunks (whether they are loaded to 60% or to 90%) depends on the size of the switching modules used to construct the switch.

Clearly, if concentration is employed, there will be times when some calls are lost. Exchanges and transmission circuits are therefore dimensioned to allow a particular level of loss in the busiest hour. This will be expressed as "1 lost call in 100", the figure often used for local exchanges, and is called the *grade of service* offered at that point in the network.

Telephone engineers use a unit to measure traffic called the *Erlang*, which is the average number of calls existing simultaneously. Expressed in this unit, the traffic figures mentioned above are:

Subscriber busy hour calling rate	0.04 E
Trunk busy hour loading, poor switching efficiency	0.6 E
Trunk busy hour loading, good switching efficiency	0.9 E

Circuit Switching and Message Switching

The descriptions so far have concentrated on the form of switching known as circuit switching. In **circuit switching** (*fig. 1.3a*), a connection is set up and maintained for the entire duration of the call. The exchanges need have no knowledge whatever of the information content being transmitted by the call. This is clearly the method most appropriate to verbal communication between two people, free to interject, comment and reply or just to remain silent, at will.

Message switching (*fig. 1.3b*), on the other hand, originated with telegraph traffic. The original message switches were boys who ran across the office with messages received at incoming terminals for onward transmission by outgoing terminal operators. Most suitable for machine communication, mes-

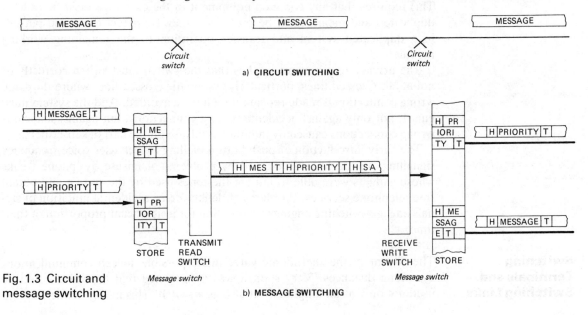

Fig. 1.3 Circuit and message switching

sage switches can store messages, broadcast them to many different destinations, and delay sending until a requested time or until sufficient messages have been collected for a particular destination. For all this to be possible, message switches must take note of certain characteristics of the information being sent. Furthermore, the terminals on a message switch cannot converse as freely as can circuit switched terminals.

The message to be transmitted over a message switching system will have to conform to a *protocol*. Typically, it will have a leading label (header) giving data about the sender, the recipient(s) and, possibly, length of message and codes and protocols to be used. It will also have a terminating label to indicate end of message with, possibly, further information to verify the accuracy of the message transmission. This message protocol is not unlike the form used for sending a telegram and has a similar purpose. Spaces in the header and end-of-message fields may be used by the message switch to add administrative data: cost, code, verification, etc.

A necessary feature of a message switch is a **store** to hold the incoming message until it has been sent completely to all the required destinations. *Figure 1.3b* shows a routine message being interrupted in order that a message of a higher priority can be sent.

Security

The requirement that a communications switch must be available to switch calls at any time introduces a requirement for a high degree of security (reliability). Another form of security, privacy, is also a prerequisite. The inventor of the step-by-step switching system, Strowger, was led to consider automatic switching because he thought his calls were being routed to a rival undertaker!

The reliability requirement for a communications switch is far more stringent than for most computer systems. A common form of the requirement is that the exchange shall not fail completely more often than once in 20 years. This requires that any common equipment in the exchange must be at least duplicated and means must be provided to always detect the failed partner in a duplicated system and transfer operation to the remaining working module.

The privacy requirement implies that the switch must switch correctly or not at all. Crossed lines, particularly systematic crossed lines where the same wrong connection is made repeatedly, must be avoided. And the system must guard, not only against accidental wrong connections, but against malicious wrong connections caused by intentional misuse on the part of subscribers.

The early introduction of push button dialling (which uses voice-frequency signalling) in the US was dogged by problems of misuse by 'phone freaks whose gadgets were able to imitate the tones used by the system and obtain free telephone services. Problems of deliberate or accidental imitation of signals engage switching engineers' attention for a significant proportion of their time.

Switching Terminals and Switching Links

The advent of the thermionic valve made possible speech communications over long distances. Very soon it was appreciated that the valve could be used not only as an amplifier but also as a switch. This in turn made possible dramatic savings in transmission links by multiplexing many conversation channels onto the one link—pair of wires, co-axial tube, radio link, etc. We therefore have to differentiate between *terminal switching*, from line or trunk to line or trunk as has been described for a circuit switched telephone exchange, and *transmission switching*, many trunks or channels or conversations on to a single link.

Transmission switching has been accomplished by *frequency division* (dividing up the available bandwidth of the link into 4 kHz sections for each conversation) or by *time division* (sampling each conversation and interposing samples from other conversations in the gaps between samples).

Switching in Space and Time

Switching implies division and already we have identified three types of switching division. *Frequency division* is used extensively in *transmission* switching where many channels (conversations) are frequency-multiplexed by modulating each conversation onto a different carrier frequency and successively modulating the resulting carriers onto higher-order carrier frequencies. *Figure 1.4* demonstrates the process of a typical frequency division transmission system.

Figure 1.4 shows how 12 analog channels are multiplexed onto a single group carrier by using the channel signal to modulate a carrier frequency unique to that channel (62 kHz for channel 1, 66 kHz for channel 2 and so on up to 106 kHz for channel 12). Group 1, consisting of twelve channels so formed, is used to modulate a 420 kHz carrier, and five such groups using five carriers between 420 kHz and 612 kHz comprise one supergroup. As shown in *fig. 1.4b*, 16 such supergroups of 60 channels are combined to give a 960-channel transmission capability.

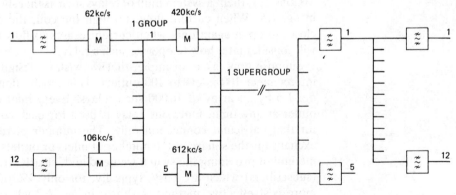

Fig. 1.4a Frequency division transmission system

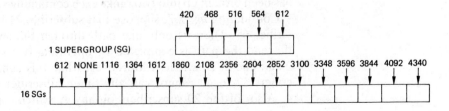

Fig. 1.4b CCITT 960-channel frequency plan

As yet, no viable system of terminal switching using frequency division has been proposed.

For *terminal* switching, therefore, we have to deal with two methods: space division and time division. **Space division**, as its name implies, provides a unique path through the exchange from inlet to outlet for each conversation. The paths are set up and held for the conversation and the switches may then be used again in a different configuration for other conversations. Clearly, space division is not an option for transmission switching since it is not able to concentrate more than one conversation on to a single transmission link.

In **time division** switching, each conversation is allocated a unique **time-slot** in a continuously cycling train of time-slots. This implies that most of the conversation is not transmitted at all, only brief, occasional samples being sent. With a suitably rapid sampling rate, the delay mechanisms inherent in the human ear are sufficient to allow the complete conversation to be reconstituted at the distant end. The time-slot allocated to the call may not actually be retained for the complete connection nor for the entire call but may be allocated anew for each sample or changed, by delay, as the sample proceeds across the exchange. Thus, time division switching does concentrate many conversations on to a few transmission links and can therefore be used for transmission as well as terminal switching.

Space Division Switching

Although Strowger's step-by-step system was the first automatic telephone switching system to be invented, it is easier to explain the principles of space division switching by first considering *co-ordinate crosspoint switching systems*. Proposed in 1912, these are still the most widespread and successful systems in general use.

Consider, then, a system built of relays such as that shown diagrammatically in *fig. 1.5*. When current flows through the coil, the contacts are made to close, to open again only when current ceases to flow through the coil. We will consider later how to operate such a relay.

Suppose now, as an example, that we wish to design an exchange to fully interconnect 100 inlets to 100 outlets. This can be done easily as shown in *fig. 1.6* by an array of 10 000 such relays. Every inlet can connect to every outlet at any time. Only one relay is used for each connection and is used for that particular connection only. The number of relays required by the exchange is the square of the number of inlets or outlets. This is an expensive method of providing the switch we need and, clearly, overgenerous for inlets (subscribers) which are busy, typically, for only 4% of the busiest hour. A more cost effective method is shown in *fig. 1.7* where the one big switch has been broken up into two ranks each containing several smaller switches.

Considering *fig. 1.7*, suppose that subscriber 21 is to be connected to subscriber 65. There is only one path through the switch which accomplishes this and the path is completed by operating two switches, one in A switch 2 and one in B switch 4. Moreover, while this connection is retained, subscribers 22 to 40 cannot call any other subscriber terminating on B switch 4. Also, of the 20 subscribers on any A switch, only a maximum of 5 can

Fig. 1.5 Electro-magnetic crosspoint relay

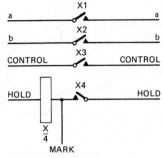

a) Full symbol

b) Abbreviated crosspoint symbol

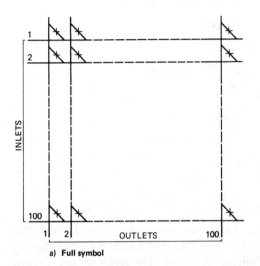

a) Full symbol

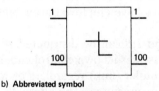

b) Abbreviated symbol

Fig. 1.6 100 × 100 crosspoint switch

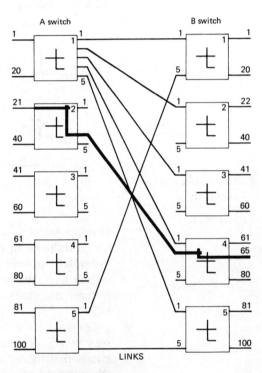

Fig. 1.7 Two-stage 100 × 100 switch: sub 21 connected to sub 65

make calls at the same time. So we have paid for the economy in relays (1000 instead of the 10 000 of *fig. 1.6*) by introducing a degree of blocking into the network; not all the subscribers can make calls at the same time.

Because of the very low calling rates of most telephone users, most telephone switching systems employ limited availability switching networks, and traffic engineering rules based on probability theory are used to dimension the network to provide the required grade of service.

The switching system shown in *fig. 1.7* consists of two ranks of switches, every switch having an *availability* of 5. We described it as being made up

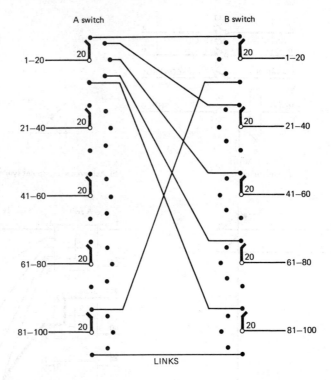

Fig. 1.8 Two-stage 100 × 100 switch using rotary switches; availability 5

of relays but it could just as easily have been constructed of rotary switches and is redrawn in this fashion in *fig. 1.8*. The traffic-carrying properties of the switch improve as the availability is increased. The Strowger *step-by-step* system uses, in the main, switches with availabilities of 10 and 20. Relay systems and other common control systems have employed availabilities up to 200.

Control of a system such as *figs 1.7* and *1.8* can be distributed as in the step-by-step system where each switch possesses its own control circuit. This means that the switch will choose a free outlet (link) regardless of whether this will lead to a busy outlet later in the connection. The alternative form of control is **common control** where the path is set from end to end of the system by a common control machine which then relases itself and can then be used to set up other calls. Common control systems will be more traffic-efficient (because they do not set up "dead end" connections) but more prone to failure. A step-by-step system is more tolerant of failure.

It is not intended to enlarge upon the subject of space division switching. It is outside the scope of this book. Sufficient has been said to introduce the concepts which are necessary in considering digital time division switching. A bibliography at the end of the chapter suggests sources for further reading. This is not the last word however. From time to time in the course of the book we shall revert to analog space-division switching in more detail, to illustrate or contrast with digital time-division switching.

Introduction to Communications Switching 11

Time Division Switching

Digital pulse-code-modulation switching is only one form of time division switching. A generalised definition of time division switching is *any form of switching which enables a common resource to be shared among a number of users in time.* Each user is allowed brief, possibly regular, "turns" during which the user has sole use of the shared resource.

The use of common control machines in space division switching was an early example of time division switching. Common stored program control of modern digital exchanges is still shared by time division in this way. Multi-user terminals to a common computer system is another form of time division. In these examples of exchange control and shared computer, the resource is dedicated to the user for a variable period, dependent upon the user's requirement and on the occurrence of a higher priority "interrupt". Time division of a shared communications link was first developed, and is still used, as a regular repetitive sharing of the common resource.

Fig. 1.9 Simple TDM switching

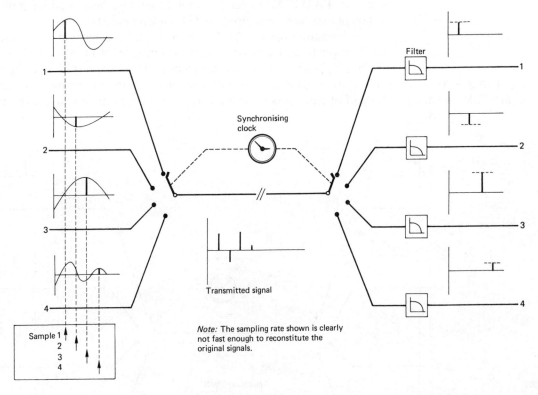

Consider first the sharing between a number of communication channels of a common transmission link (*fig. 1.9*). At each end of the link is a continuously rotating switch and the switches are synchronised so that they rotate at the same speed. Each incoming channel is sampled in turn and the message over the link consists of a series of samples, one from each channel. Provided the switches remain synchronised, the channels are unscrambled from the common message at the receiving end and allocated to the correct receiving channel. Provided the sampling rate is fast enough, then the delay mechanisms of the human ear and the transmission line properties of the channels will restore a reasonable representation of the original unsampled message.

A theory called Nyquist's sampling theory shows that to retain the integrity of the input signal the sampling rate must be at least twice the frequency of the highest frequency signal to be transmitted. For speech telephony, whose bandwidth is 300 Hz to 3400 Hz, a sampling rate of 8 kHz is therefore adequate. This represents a sample of the same channel once every 125 μs. *Figure 1.9* shows a sampling rate which is too slow to reconstitute the original signal but a realistic diagram would be difficult to follow.

The system described represents an example of pulse amplitude modulation since the sample pulse of signal is sent to line as it stands, and the height of the pulse indicates the nature of the input signal when sampled. Clearly, such pulses will be degraded on transmission and are just as difficult (or even more difficult) to reconstitute by amplification as they pass down the transmission line. PAM TDM systems have not therefore been used for transmission although they have been used for TDM telephone switching.

The breakthrough in TDM switching and transmission was indicated in the Introduction: it was Alex Reeve's invention of Pulse Code Modulation. If the amplitude sample can be measured and expressed as a digital code before transmission, then problems of noise and attenuation are removed. Instead of amplifiers we can use repeaters which merely detect the presence

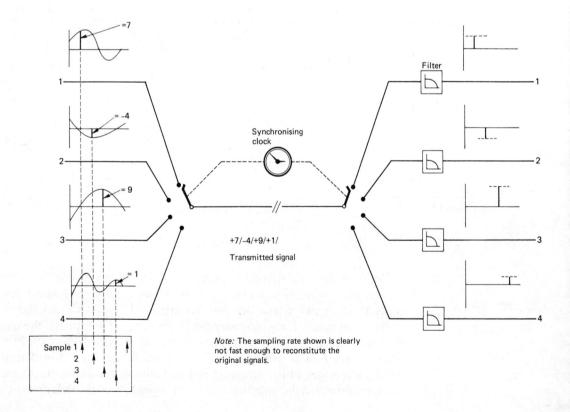

Fig. 1.10 Simple digital TDM switching

Introduction to Communications Switching   13

or absence of 1s (ones) and re-transmit new 1s to the distant end.* *Figure 1.10* illustrates our simple TDM system modified in this way.

The circuitry involved in converting amplitude samples into digital (binary) code and transmitting the code to line within 125 μs is not simple and would have required roomfuls of thermionic valves. Further rooms full of equipment would be needed to provide them with sufficient power and equal or greater volumes of equipment would be necessary to keep them cool. We had to wait, therefore, for the transistor and succeeding stages of integration before practical PCM transmission systems and, later, switching systems became a reality.

Signalling, Routing and Control

It is now time to consider again the activities of the operator at a manual board. These activities are listed in Table 1.1 and categorised by function and by the phase of the call during which they take place.

Table 1.1 Call Control Activities

Activity	Function	Call phase
Look for new calls	Scanning/Signalling	Pre-selection
Determine user wishes	Signalling	Pre-selection
Determine exchange routing	Routing	Pre-selection
Determine network routing	Routing	Pre-selection
Operate exchange switches	Operation (marking)	Call completion
Seize trunks	Signalling	Call completion
Send on call details	Signalling	Call completion
Ring called user	Signalling	Call completion
Recognise answer	Signalling	Call completion
Connect	Operation	Call completion
Start charging	Operation (charging)	Call completion
Supervise call	Scanning	Conversation
Recognise release	Scanning/signalling	Release
Disconnect	Operation	Release
Complete charge record	Operation (charging)	Release

Table 1.1 divides the activities related to a call into phases as follows:

Pre-selection   All those activities related to recognising a new call request and determining how to deal with it.

Call completion   The actual connection of the call and starting the charging process.

Conversation   The stable period of the call during which the parties communicate.

Release   Disconnection of the call, completion of the bill, and restoring the network to the normal (idle) state.

* This assumes, of course, that the digital information is in binary form, not decimal as shown in *fig. 1.10*.

Table 1.1 also identifies the various functions performed during the activities related to the phases of a call. These are:

Scanning Looking for new calling conditions, also looking for free switch paths and free trunk circuits. Looking for free paths and trunks is also known as *interrogation* or *search*. Scanning may also be involved in looking for release of an existing call during the conversation phase (supervision).

Signalling A calling condition is a signal; signalling is a quite wide-ranging function of generating (sending) and recognising (receiving) signals. We can also distinguish three types of signals: those related to the connection itself (seize, answer, release) known as line signals, those related to choosing the correct connection (proceed to send, digits, end of message) known as register signals, and those related to the machine instructions to the user.

Routing All those activities related to converting the users' wishes into particular connection commands. Within the exchange it will be a relatively simple procedure of choosing the "best" free path. This will involve translating the user's instruction (say 857679) into "set A switch nn, B switch pp, C switch qq, etc.". Within the network it will be a quite complex process of converting the user's request into routing instructions for the originating exchange and routing instructions to be sent on to control transit exchanges and the called user's terminating exchange.

Operation The actual actions of setting-up or breaking-down the paths.

Charging Charging is one example of a whole host of tasks which the exchange must perform on behalf of the administration. These tasks also include reporting faults, recording traffic, responding to changes in operating procedures, etc.

To understand these functions in a little more depth we will illustrate each main function, signalling, routing and control, with examples. Scanning, operation and charging can all be considered as parts of the overall control function.

Signalling

Part of Strowger's original concept for a step-by-step automatic switching system was the idea that the user's request should be sent to the exchange as a series of alternate loop disconnections and connections. Each disconnection would signify one digit. This concept is only now being replaced by modern push-button telephones. The process of setting-up a call through a step-by-step exchange is illustrated in *fig. 1.11*. Some time spent studying this diagram will afford a basic understanding of not only the principles of step-by-step exchanges, but also the three forms of signalling met with in communications switching. These three forms of signalling are:

Introduction to Communications Switching 15

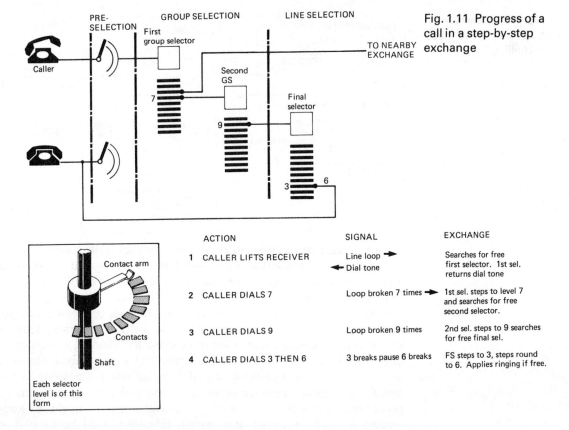

Fig. 1.11 Progress of a call in a step-by-step exchange

	ACTION	SIGNAL	EXCHANGE
1	CALLER LIFTS RECEIVER	Line loop → ← Dial tone	Searches for free first selector. 1st sel. returns dial tone
2	CALLER DIALS 7	Loop broken 7 times →	1st sel. steps to level 7 and searches for free second selector.
3	CALLER DIALS 9	Loop broken 9 times	2nd sel. steps to 9 searches for free final sel.
4	CALLER DIALS 3 THEN 6	3 breaks pause 6 breaks	FS steps to 3, steps round to 6. Applies ringing if free.

Each selector level is of this form

User signals In this example, dial tone and ringing. Instructions and information from the machine to the human user.

Line signals In this case, seizure (loop) and clear (long disconnection of loop).

Routing signals In this case, trains of short breaks interspersed with longer replacement of the calling loop. The telephone dial mechanism inserts the longer replacement of the loop automatically as it returns to its home position. Thus digit 3 is represented by 3 break/make impulses followed by a longer make.

Had the caller in *fig. 1.11* dialled 8 instead of 7 the call would have been routed over a trunk to a neighbouring exchange. In this case the remaining digits would have had to be repeated over the trunk to set up the connection in the terminating exchange. This simple system of loop disconnect signalling is summarised in *fig. 1.12*.

In all common control systems and in some step-by-step systems, notably the director system used in the large UK conurbations, a device called a

Fig. 1.12 Minimum basic loop disconnect signalling

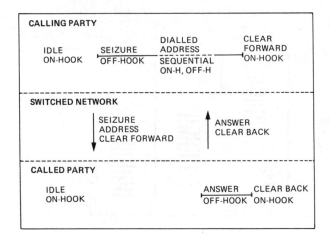

register is used to receive and remember the user instructions. The register can apply for a translation which will convert the dialled information into a different series of digits more suited to the network. For this reason the routing signals are more often called register signals.

Until the very recent past, the signalling channel between exchanges was the very same channel as that used for the voice connection. This seemed so self-evident that the term used to describe it, "channel associated signalling", was not coined until an alternative method was envisaged. The introduction of common control using stored program made it possible to propose that the telephone exchange network could profit from allowing the control processors to communicate directly. Control functions might no longer be located at each exchange, but certain functions could be located at one exchange but be used by the simpler processors at several neighbouring exchanges.

To achieve this degree of inter-processor communication requires a data link between processors which is secure and of high capacity. Why should this same data link not be used for the signalling requirements of the individual voice circuits? This concept of **common channel signalling** (CCS), illustrated in *fig. 1.13*, greatly simplifies the trunk termination equipment for the individual trunks at each exchange and utilises more efficiently the high-capacity secure communications link provided for inter-processor communication. The concept is now generally accepted and is already in use on international circuits and in the US national network. CCS is the preferred signalling method on the emerging digital network.

Fig. 1.13 Principle of common channel signalling

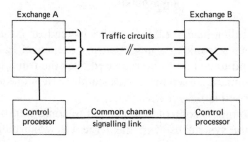

The principles of CCS are described in some detail in Chapter 7. Indeed, as this book is concerned with digital communications switching, more space is devoted to CCS than to channel associated signalling. Nevertheless, the criteria and concepts governing signalling communications relationships between exchanges have been developed in an environment dominated by channel associated signalling. This dominating influence will remain a factor in defining the communication between digital SPC exchanges over CCS links for a long time in the future.

Routing

In the example of *fig. 1.11* we see that the network routing information is inherent in the way in which the various switches are connected together. First selector level 7 always connects to local calls, level 8 always connects to the neighbouring exchange. In order that users connected to the neighbouring exchange can have a common directory with those on *fig. 1.11* exchange, their first selector level 7 must connect to *fig. 1.11* exchange. Clearly, for a national network of thousands of exchanges this is an intolerable constraint and register/translation systems are utilised to divorce the national (and international) numbering scheme from the network routing.

Suppose, for example, a London user calls a Bristol number. The London user dials (keys) the national number: 0272 857679. The decisions taken by the exchanges involved in the call are as follows:

0   Originating exchange recognises that it is a national subscriber trunk dialling (STD) call. The call is routed to the parent *group switching centre* (GSC) where the required library of translations is located.

272   GSC converts this to the routing code to reach Bristol GSC from the originating GSC. Note that this routing may vary from every originating GSC although the dialled code is identical.

85   Bristol GSC recognises this as the code for Nailsea exchange.

7679   Nailsea exchange connects to subscriber 7679.

The GSC in London may have various routes available to Bristol and will choose the most suitable route according to rules pre-recorded in the exchange data memory. Similarly, the Bristol GSC may have to choose a route to Nailsea.

Control

All the functions of signalling and routing described so far interwork with the exchange control function to provide a complete capability. Included in control are the testing of the exchange network for free paths and of the trunk network for free trunks. Control must also busy any paths and trunks chosen. Control also performs the actual operation of setting up the connection. We showed a crosspoint relay in *fig. 1.5* and a matrix of such relays in *fig. 1.6*. In *fig. 1.14* we illustrate how control marks a path through a network of such relay matrices and actually operates them.

18 Introduction to Digital Communications Switching

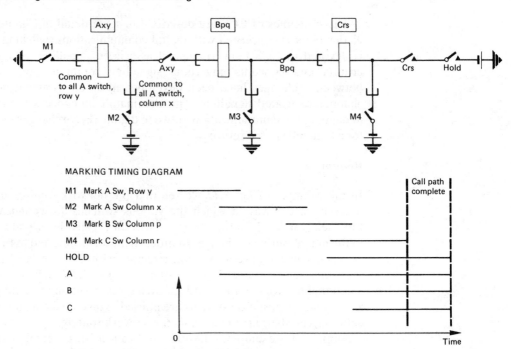

Fig. 1.14 Marking of a three-stage coordinate switch

Control also performs all the charging functions and other administration functions and must further perform its own housekeeping functions, particularly those related to recognising when it is making mistakes and handing over control to a standby machine.

Chapter Summary

We have completed a rapid tour through the subject matter of the whole book. Having defined the objective of communications switching as connectability at any time, we considered the concepts of communications traffic, the methods of switching whether by circuit or by message, and the need for security. We then considered in more detail the switching tasks of switching between terminals and switching (multiplexing) over links. Concentrating then on terminal switching, we outlined the methods of space division and time division switching. Finally, we considered the functions of control required to achieve automatic switching, dwelling at some length on the signalling and routing functions.

References (Introduction and Chapter 1)

[1.1] UK Patent Specification 535860, Application date, 23 October 1939.
[1.2] Shannon C E, A Mathematical Theory of Communication, *Bell System Technical Journal*, Vol. XXVII, No. 3, July 1948.
[1.3] Povey P J, *The Telephone and the Exchange*, Pitman 1979.
[1.4] Leakey D M, *Switching in Space and Time*, Paper 7781E, IEE Electronics Division, October 1976.
[1.5] Robertson J H, *The Story of the Telephone*, 1947.

Exercises 1

1.1 Re-draw the network of *figs 1.7* and *1.8* using rotary switches with an availability of 20. How many subscribers connected to the same rank of A switches can call at once and how many other terminals are blocked from each other by any A switch to B switch link being in use?

1.2 Figure 1.13 is a very condensed explanation of a typical method of marking a path through a three-stage relay switching array. Write out a more complete explanation in the form:

    M1 operating   marks with earth all A switch relays in row y.
    M2 operating   marks with −50 V all A switch relays in column x. Only relay Axy is backed by an earth from M1 and this relay alone operates.

As you continue be sure to explain why relay Bpq does not operate immediately M3 operates.

1.3 A co-ordinate switching array has 12 A switches of 16 inlets and 12 outlets to links to the B switch of the array.
    a) How many B switches will there be in the array and how many inlets to each B switch?
    b) If the A switch inlets each carry 0.1 E, what will be the loading in erlang of each A-B link

1.4 Write down, in the form of a list of advantages and disadvantages, the relative merits of PAM TDM transmission and PCM TDM transmission.

1.5 Categorise into the phases of a call the following call events:
    Send proceed to send signal
    Recognise clear forward signal
    Start charging
    Mark connection path

1.6 Classify the following signals as user signals, line signals or routing (register) signals:
    End of sending
    Backward clear
    Dial tone
    Meter pulse
    Seize pulse
    Inter digit pause (signal)

2 Introduction to Communications Transmission

The mathematical treatment of the behaviour of electrical signals launched down transmission lines was available almost before many such signals had been sent. (This is, perhaps, the only portion of a syllabus which has remained the same for succeeding generations of engineering students.)

Our interests in communications transmission are centred upon the media and methods used, in particular where these affect the switching equipment and where digital techniques have brought about a combination of transmission and switching.

Wire Transmission

Historically the first form of transmission link was the metallic conductor line. Early telegraph circuits were realised by one-wire earth-return lines. Differences in earth potential over long telegraph circuits caused noise and sensitivity problems and led to the use of two-wire circuits. Such circuits are fundamentally "one-way" and require switching means or a bridge network to render them suitable for both-way operation. The terms used for one-way and both-way operation are *simplex* and *duplex* respectively. To make a truly **duplex transmission link** requires four wires (or two circuits of one wire and earth return at least). Clearly also a four-wire circuit is necessary if amplification is to be used in the circuit.

A four-wire circuit or a number of two-wire circuits introduce a further link, known as the **phantom**. This emerges through using the wire already in use in one half of a duplex connection as one wire in a further simplex connection. The phantom circuit can be used as signalling channel or as an additional transmission channel. All these arrangements are illustrated in *fig. 2.1*.

Wire Carrier Telephony

Use of the phantom provided a maximum of three communication channels for every two actual physical circuits. Any greater packing density requires the use of some form of multiplexing technique. A simple form of 12-channel carrier technique is used to illustrate one method of achieving this using frequency division multiplex (*fig. 2.2*).

Each channel incoming to the system is limited by filters to the frequencies between 300 Hz and 3400 Hz. The signal of each channel is used to modulate a carrier frequency unique to that channel and the 12 modulated carrier frequencies are sent to the single line. At the distant end a bank of 12 filters, tuned one to each of the carrier frequencies, extract the individual channels, and the modulating signal is sent to the output channel terminal.

Introduction to Communications Transmission 21

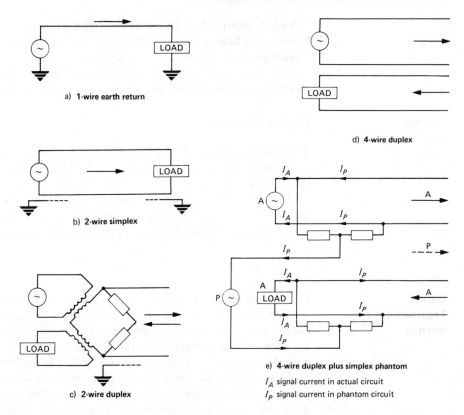

Fig. 2.1 Wire transmission modes

Fig. 2.2 Wire-carrier telephony principle

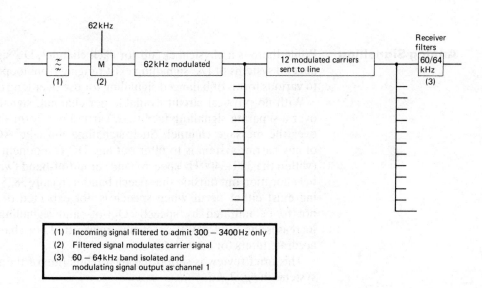

(1) Incoming signal filtered to admit 300 – 3400 Hz only
(2) Filtered signal modulates carrier signal
(3) 60 – 64 kHz band isolated and modulating signal output as channel 1

Such a group of 12 channels multiplexed on to a single carrier circuit is known as a **basic group**. Further combinations of five such groups produces a **supergroup** of 60 channels.

(The description above and *fig. 2.2* neglect altogether the identical process which must occur for the opposite direction of transmission.)

Carrier Telephony

The same principles of carrier multiplexing are used for multiplexed telephony on co-axial cables, radio systems, etc. In each instance there will be detailed differences because of the medium. *Figure 1.4* showed the scheme for the CCITT 960 channel co-axial cable plan and *fig. 2.3b* shows the frequency allocation for a submarine cable system. *Figure 2.3a* is a block diagram of a repeater in sufficient detail to explain some of the allocations of *fig. 2.3b*.

Other Transmission Media

Co-axial cable has already been mentioned as a high-capacity transmission medium. A further commonly used medium is *microwave radio* which is particularly useful in sparsely populated areas, difficult terrain and areas where the value of the copper encourages dishonest people to purloin the cables. Telephony circuits are also completed (decreasingly frequently at the present time) over *HF radio links* and using *tropospheric scatter radio systems*. These latter are still in frequent use on marine oil installations. Lastly, *satellite systems* are viable for intercontinental links and presently share most such connections with submarine cable systems. Increasingly, satellite systems will become useful for those national links presently served by microwave radio although the satellite delay of about 0.5 sec per hop limits the possibilities of multi-satellite connections.

Analog Signalling

While there is a physical circuit for each channel, DC signalling can be used. Several systems of DC signalling exist ranging from loop disconnect signalling to various forms of balanced signalling for use over long distances.

With no physical circuit available per channel, signalling has to be either over a separate signalling facility or carried in a form similar to the message over the message channel. Such signalling must be AC (since the first act of any carrier system is to filter out any DC component) and can be **in-band** (within the 300–3400 Hz speech band) or **out-of-band** (within the carrier channel allocation but outside the speech band, typically 3825 Hz). In-band signalling must either occur when speech is not expected or be different enough not to be imitated by speech. Out-of-band signalling, if at all complex, increases the total power spectrum transmitted per channel and may require accurate filters for each channel.

This brief review is supported by a tabulation of the varieties of signalling systems in *fig. 2.4*.

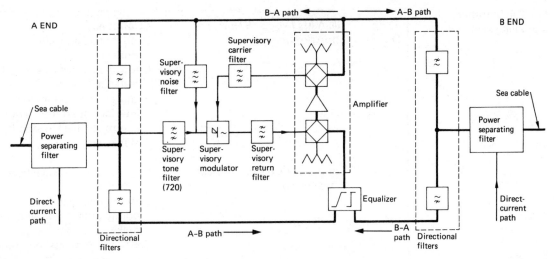

Fig. 2.3a Simplified block diagram of 120/160-channel submarine repeater

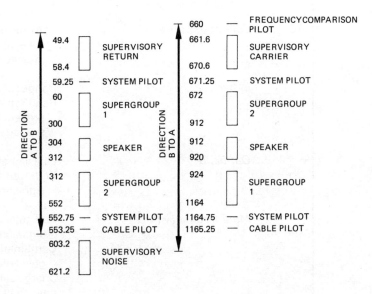

Fig. 2.3b 120/160-channel submarine system frequency spectrum

Transmission Quality

The quality of the speech transmission can be impaired in a number of ways:
　Loss of signal due to transmission impedance.
　Distortion of signal due to imbalance of the transmission impedance.
　Addition of signal by noise, inherent in the transmission channel, and by crosstalk, intelligible noise picked up from other channels.

Loss　The basic resistive and capacitive constants of the transmission line will constitute a certain loss. The overall loss of a total connection is of interest, but to allow full world-wide connectivity, the loss contributions of the various sections of a connection must be defined and apportioned.

Fig. 2.4 Summary of signalling systems

TYPE	SIGNALLING SYSTEM	COMMENTS
DC	Loop disconnect	Used for line and selection signals. Good and simple over short distances. Requires no incoming end interface apart from switch control, at least in step-by-step systems.
	Long-distance symmetrical waveform	Used for line and selection signals. Sending impedance is equal during pulse break and pulse make. Sensitive. Requires incoming interface.
AC	Continuous tone	Tone off idle (on during speech) is not possible with in-band signalling.
	In-band	Danger of speech imitation. Requirement to split line to prevent link to link propagation of signals.
	Out-band	Line signals only. Preferred system for simple signalling of few conditions.
	Multi-frequency (MF)	Particular version of in-band signalling. Combinations of pairs of tones give voice immunity and large signal repertoire. Inter-register signalling.
	MF Versions CCITT R1	Forward signals only. Standard inter-register system in US.
	CCITT R2	Compelled signals. World inter-register system standard.
	SSMF 6	Semi-compelled. UK inter-register system standard. Similar to R2.

Distortion The absence of inductance terms from the characteristic impedance of telephone cables is a prime source of distortion. The problem was solved at an early stage in the history of telephony by inserting "loading coils", adding inductance artificially at intervals along the cable. The loading coils remained and were incorporated in the carrier telephony system specifications but had to be removed for digital telephony. The loading coil spacing was one of the parameters determining the sizing of the US and early UK 24-channel digital systems.

Any transmission line, however well designed and loaded as necessary, will contribute to distortion because of the variety of terminating impedances switched to the line at either end. The inevitable mismatching of terminating impedances and line impedances will result in reflection distortions. It is not possible, using automatic switching, to adjust the parameters of the particular terminating circuit to those of the line.

Noise Noise is introduced by the speaker (background noise of the office) and at every stage of the connection. Noise introduced in the transmission link is made less important by a technique called **companding** (fig. 2.5).

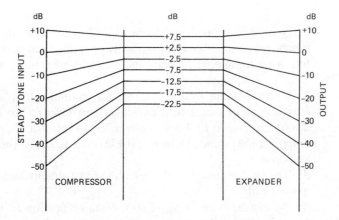

Fig. 2.5 Compandor function

Matched compressor at the sending end, and expandor at the receiving end, apply greater emphasis to low-level speech components over the transmission link than the higher levels. Thus, provided the noise level introduced by the link is low it will be suppressed compared to signal. Inherent high levels of noise fed to a companded link may actually be made worse by the use of companding.

Crosstalk The use of multiplexing techniques in itself introduces the possibility of intelligible break-through from one conversation to another. The large quantities of wiring and switches in the telephone exchanges will also introduce crosstalk, some intelligible, some at least recognisable as switching noise, etc. Security demands that crosstalk be reduced to unintelligible levels and transmission quality requires a smaller limit still.

Transmission Quality and Network Planning

Because the automatic telephone system is now an international machine, the allocation of the various components of transmission quality are subject to international regulation and control via the CCITT. It is therefore necessary for each nation to conform to certain standards in its national network in order that international calls will not experience unacceptable standards.

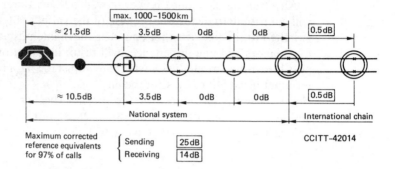

Fig. 2.6 Examples of the allocation of maximum CREs for an international connection in an average-sized country

Figure 2.6, reproduced from CCITT recommendation G121, illustrates a typical recommended international apportionment of transmission losses. The following comments are relevant in considering this apportionment:

1 Every feeding bridge, where, for example, loop signalling is separated from speech in a telephone exchange, introduces a loss of about 1 dB.

2 At the 2-wire to 4-wire conversion point of a connection which includes amplification, some loss must exist in order to adjust the circuit for best overall response.

3 The bridge circuit of the telephone instrument itself introduces some loss, at least 1 dB.

4 As can be seen in *fig. 2.6*, the major portion of the loss is specified to occur in the local portion of the connection. It is important that this should be so as otherwise there might be substantial differences in performance between local, national trunk and international connections.

5 *Figure 2.6* shows the local connection terminating on a telephone. In many cases the termination may be a PABX with perhaps the possibility of onward connection over a private network. In such cases the private network operator and the public network administration must come to some arrangement over the allocation of the overall loss "budget" for the complete local connection.

In Perspective: the Relationship between Transmission and Switching

Much has been said so far about multiplexing to combine many channels onto a single transmission medium. The place of the transmission links in the overall connection path has been discussed in terms of transmission quality. It is now appropriate to look at the overall connection, subscriber to subscriber, and this is shown, for a typical connection, in *fig. 2.7*.

Introduction to Communications Transmission 27

The figure shows switching, per call, from subs line to trunk and switching (multiplexing), on a semi-permanent basis from trunk termination to channel. The exchanges do not know, nor do they need to know, that trunk r, trunk route Y is not a physical circuit but a particular channel of a FDM system.

In fact the universal switching capability of the exchanges depends on the network losses being apportioned to a plan. Were this not the case the exchanges might have to be barred from completing certain connections.

By contrast to *fig. 2.7*, *fig. 2.8* shows a similar connection completed in a network where all trunks and all exchanges are digital. Not only has the separate multiplex switching disappeared (it is now performed as part of the exchange switching function) but all switching actions have simplified to the same type: "Switch 30 channel system X channel y to 30 channel system R channel s". Furthermore an end-to-end digital connection is effectively

Fig. 2.7 Progress of a call: analog environment

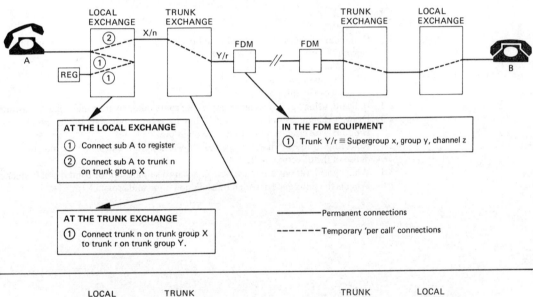

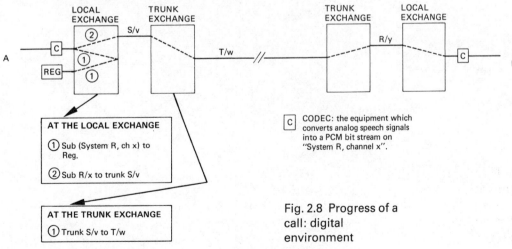

Fig. 2.8 Progress of a call: digital environment

loss-free. If regeneration is provided sufficiently often to re-constitute the original signal before its simple on/off meaning is so distorted as to be lost entirely, the digital signal received at the far end will be identical to the digital signal sent. Thus the only losses and distortions will be those introduced by the coding and decoding processes. Digital crosstalk, were it to exist, would produce a meaningless bit stream.

Chapter Summary

We have introduced the necessary minimum to understand the references to transmission, transmission systems and transmission qualities elsewhere in the book. The knowledge so gained has been used to contrast the transmission environment using analog and digital switching.

Exercises 2

2.1 Comment on the sources of noise and distortion in the following wire transmission systems (see *fig. 2.1*):
    a One-wire earth-return.
    b Two-wire simplex.
    c Two-wire duplex.

2.2 Indicate, using *fig. 2.1c* as an example, the route taken by transmission echo distortions.

2.3 Draw a diagram similar to *fig. 2.1e* but including both ends of the circuit and showing two four-wire circuits plus a four-wire phantom.

2.4 Referring to the submarine repeater shown in *fig. 2.3a* describe the function and method of use of the directional filters.

2.5 What considerations are relevant in choosing a suitable out-of-band signal frequency?

2.6 What is the difference between transmission noise and crosstalk?

3 Pulse Code Modulation

It is now time to consider the process of PCM encoding and multiplexing in detail. Historically, PCM has been introduced to improve transmission characteristics and in practice it is, above all, a transmission system. The fact that switching can be performed also in PCM is, as it were, an added bonus. This chapter, therefore, is devoted entirely to PCM transmission but, because PCM can be, and is being, used for switching, the ideas presented in this chapter are fundamental to the whole book.

The principle embodied in pulse code modulation is *to obtain a time-divided sample of an analog signal, encode each sample into digital form, and transmit a digital bit stream representing the numeric values of the series of encoded samples*. The process is reversed at the receiving end: the signal is demultiplexed, de-coded and integrated to regain the analog signal. Thus the signal is transmitted in a series of simple "there" or "not-there" digital signals easy to regenerate at intervals and impervious to noise and distortion introduced by the transmission medium.

The easiest way to understand PCM is to consider how it works using a series of diagrams. In this chapter we will consider the 30-channel (the World standard) and the 24-channel (the American standard) systems in some detail. For explanation purposes we will limit our pictorial representation to a 4-channel system; this is called the Example System.

Example PCM System

Figure 3.1 shows a part of the four signals of the four channels. It can be seen that all four channels are active; there is a signal present on each channel. The horizontal axis of the graph is the time axis but is marked off in channel numbers as well as in milliseconds.

Sampling *Figure 3.2* shows the sampling process for each of the four channels. Channel 1 is sampled at 0, 12, 27, 39, etc. milliseconds (ms). Channel 2 is sampled at 3, 15, 30, 42, etc. ms. The result of the sampling process for, say, Channel 3, is a pulse of amplitude 2.5 units at 6 ms, a pulse of amplitude minus 1 unit at 18 ms, and a pulse of minus 2.9 units at 33 ms.

There is a puzzling discontinuity in this timing, which will be explained.

Fig. 3.1

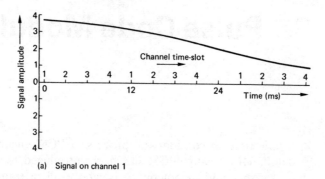

(a) Signal on channel 1

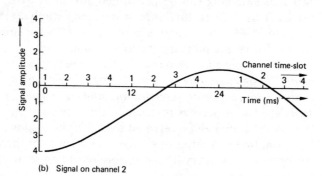

(b) Signal on channel 2

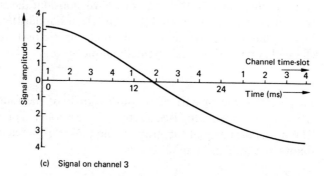

(c) Signal on channel 3

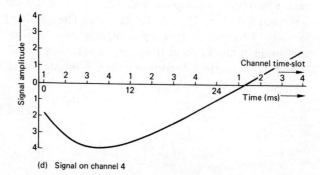

(d) Signal on channel 4

Fig. 3.2

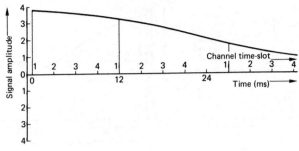

(a) Signal on channel 1: sampled

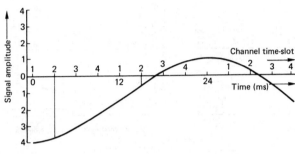

(b) Signal on channel 2: sampled

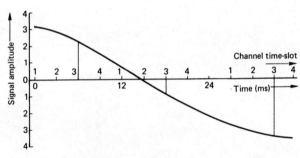

(c) Signal on channel 3: sampled

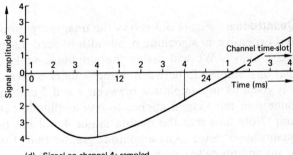

(d) Signal on channel 4: sampled

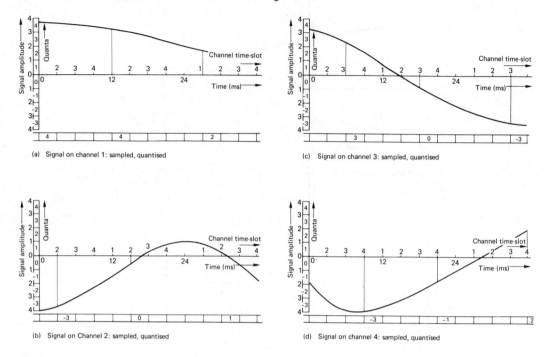

Fig. 3.3

Quantisation *Figure 3.3* shows the quantising process. It would be difficult and expensive in signalling bandwidth to send a signal: "the sample at 6 ms is 2.5 units". We will divide the amplitude scale into bands or **quanta** so that any amplitude between 3 and 4 units is given the value 4. The value 3 is given to any amplitude between 2 and 3 units and so on. Note that the value 0 in this system applies to any amplitude between zero and minus 1 unit. Note also that the quanta minus 4 will not be used; it is outside the quantisation range. Any amplitude greater than 4 units will be quantised as 4; any amplitude less than minus 4 units will be quantised as minus 3.

Pulse Code Modulation

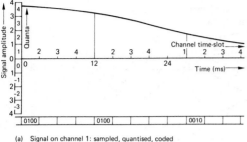

(a) Signal on channel 1: sampled, quantised, coded

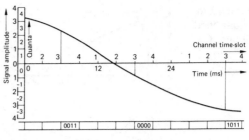

(c) Signal on channel 3: sampled, quantised coded

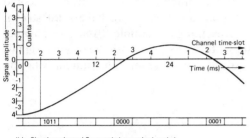

(b) Signal on channel 2: sampled, quantised, coded

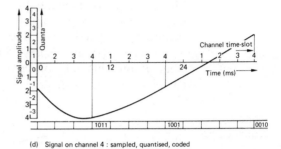

(d) Signal on channel 4: sampled, quantised, coded

Fig. 3.4

Coding So far, then, we have converted the transmission requirement from continuous sending over four separate channels to sending one of eight quanta signals per channel in time-divided time-slots. Eight different signals is still rather complex so the coding process of *fig. 3.4* will be employed. The example will be coded in 3-bit binary plus a sign bit. The signal repertoire is, therefore, as shown in Table 3.1.

Table 3.1 Example System Coding Structure

Signal Content	Binary Meaning	Decimal Meaning
0111	Not used	Not used
0110	Not used	Not used
0101	Not used	Not used
0100	+100	+4
0011	+011	+3
0010	+010	+2
0001	+001	+1
0000	000	0
1001	−001	−1
1010	−010	−2
1011	−011	−3
1100	Not used	Not used
1101	Not used	Not used
1110	Not used	Not used
1111	Frame alignment	Frame alignment

Fig. 3.5 Channels 1 to 4: sampled, quantised, coded, multiplexed

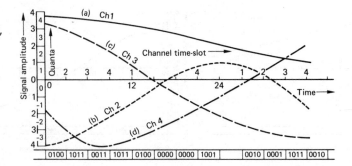

Multiplexing By now the presentation adopted in the diagrams has become unnecessary. Having sampled, quantised and coded, we have a **bit stream** prepared, ready to be sent to line representing the samples from each channel in time divided order. *Figure 3.5* shows this bit stream with the original signals from all four channels collected onto the one diagram, thus summarising the complete process so far.

Fig. 3.6 Channels 1 to 4: sampled, quantised, coded, multiplexed; frame bit stream sent to line

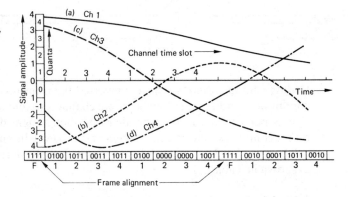

Alignment A difficulty remains however: how is the receiving end to know which is Channel 1, which is Channel 2, etc.? The time divided series of messages for 4 channels is a frame and for the example system we have inserted a **frame-alignment signal** as an additional 4 bits sent once every 2 frames. (It was in preparation for this that earlier diagrams left a space for the frame-alignment signal.) Because we have chosen to frame-align every second frame, the sampling rhythm will be slightly irregular as shown in *fig. 3.8*. The receiving end will be expected to remember (or free-wheel) to stay in time with the transmitting end over the intervening frames between alignment time-slots. Addition of frame alignment is shown in *fig. 3.6*.

Reception Lastly, at the receiving end, the message is received, de-multiplexed, de-coded and integrated, resulting in the output analog signal shown in *fig. 3.7* for Channel 1 only. The receiving process could have been illustrated in the same detail as *figs 3.1* to *3.4* showed the transmitting process. (Reproduction of elements of the receiving process is left as part of the Chapter exercises.)

Fig. 3.7 Channel 1 received, demultiplexed, decoded, integrated

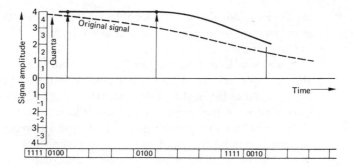

Discussion of Sample System

The PCM process has been described using the example of an entirely imaginary system. Before leaving this system and moving on to practical PCM systems we will summarise the qualities of the imaginary system and use it to highlight some of the difficulties that practical systems employing rather more sophistication need to overcome. The sample system used had, in summary, the characteristics shown in Table 3.2. It is summarised again in *fig. 3.9* in a format which will be used to describe the practical systems.

Table 3.2 Example PCM System: Summary Data

Frequency range	10 (say) to 37 Hz (see *Note 1*)
Sampling rate	74.07 Hz
Bits per sample	4
Time-slots per frame	4.5
Channels per frame	4
Output bit rate	1333 bits/sec (see *Note 2*)
Encoding law	linear
Signalling capacity	none

Note 1 Sampling theory indicates that the maximum frequency transmitted will be half the sampling frequency. Our sampling period is the mean of 12 and 15 ms or 13.5 ms. Hence:

$$\text{Maximum frequency} = \tfrac{1}{2} \times \frac{10^3}{13.5} = 37 \text{ Hz}$$

Note 2 Output bit rate $= (4 \times 4 + 2)\dfrac{10^3}{13.5} = 1333.33$ bits/sec

Considering Table 3.2 it is apparent that the bandwidth used, effectively the same as the output bit rate of 1333 bits/sec, is large compared to the tiny frequency range transmitted. 10 to 37 Hz is just enough to send a digital version of the telephone ringing signal. It is because PCM is so greedy in bandwidth that its introduction had to await cheap and freely available switching devices. We also see from Table 3.2 that the encoding law is linear. There is no attempt to introduce the digital equivalent of the compandor discussed in Chapter 2.

If we now look back at *figs 3.1* to *3.9* we can see further drawbacks of the example system. Clearly, 4 channels is not sufficient; a basic modular unit at least as big as the FDM systems, namely 12, would be more desirable. Comparison of the original signal and the received signal (*fig. 3.1a* and *fig. 3.7*) shows that many more quantising steps are needed. Also, if the quanta are made smaller for small signals, thus coding small amplitudes more accurately than large, then an effect similar to the compandor will be achieved.

Considering coding, even this arbitrary example includes long series of zeros and ones. After a series of zeros or ones has persisted for the space of little more than a time-slot, the signal sent to line ceases to be a true alternating signal. In any case, unless the balance of zeros and ones is roughly equal, the frequency sent to line will be very variable in an arbitrary way. There is therefore a need to modify the signal in some way to make it continuously alternating and to maintain a substantially constant frequency.

The example system had to include an additional time-slot to provide frame alignment as a means of keeping the receiving end in step with the transmitting end. What the example system does not have is any provision for signalling information either related to the individual channels or relating to the PCM link as a whole. A practical PCM system will have to include signalling means for both these purposes. In fact, our example system does include the inherent possibility of in-band signalling within the transmission band because such signals would be encoded and decoded like any other analog signal within the band. This would require the provision of signalling equipment per channel as in analog systems. It is preferable to utilise the opportunity for common signalling offered by the introduction of PCM. Further, signalling has lower requirements for distortion and speed and does not therefore have to be sampled as frequently as voice. Advantage can be taken of this feature in providing signalling over PCM. For practical systems we would wish to be free to use any signalling system already in use in analog systems and, if possible, to offer improvements on these for wholly digital systems.

Lastly, the "hop one, skip one" rhythm of the example system adds unnecessary complexity. It suggests that the system could be smoothed by inserting a dummy time-slot in the alternate frames and that this time-slot could usefully be used for signalling.

Practical PCM systems eliminate or greatly reduce all these problems, as shown in the next section. Having dealt with these, two further problems remain for discussion: line codes, which will form the final part of the present chapter, and quantisation, which will be dealt with in full in the next chapter.

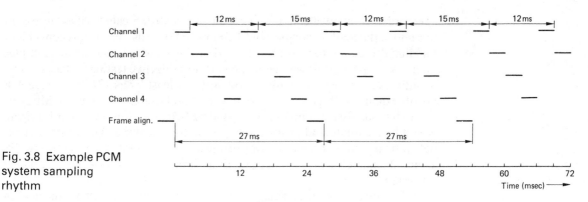

Fig. 3.8 Example PCM system sampling rhythm

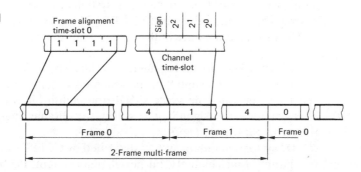

Fig. 3.9 Example PCM system

Practical PCM Systems

So it is now time to introduce practical PCM systems and demonstrate how these systems overcome the difficulties and to identify the remaining problems that practical systems still present to the administration, the manufacturer and the user.

There are some characteristics of a practical PCM system that are fixed immediately without further consideration. These stem from the internationally agreed transmission standard for voice of 300–3400 Hz and the information theory requirement that the sampling rate be twice the highest frequency transmitted. To enable the international 4 kHz band to be carried, thus including outband signals, the sampling rate must be 8 kHz. Hence, for all practical systems we have:

    Audio frequency band   300–3400 Hz
    Sampling rate   8000 Hz

Historical Summary

The first practical PCM system was introduced by AT&T in the United States. It was known originally as the T1 system but was subsequently renamed and is now more usually referred to as the D1 system. Considerations used in developing the T1 system have, to an extent, influenced the design of all practical systems.

In order to allow PCM transmission over existing cables, it is necessary to remove the loading coils spaced 1.83 km (2000 yards) apart. It is convenient for digital regenerators to be provided at the same spacing (or at multiples of the same spacing). It was determined from subjective testing that acceptable quantisation distortion required coding to at least 7 bits (7 bits provide 128 possible quanta). Regenerator spacing for digital streams of about 1.6 Mbit/sec would be roughly 1.83 km suggesting that a 24-channel 8-bit-per-sample system would be adequate and provide some room for signalling. Twenty-four channels also matches the modularity of most carrier FDM systems. In the US some 12-channel rural systems would be directly replaced by the 24-channel T1 system.

Although this book is going to be concerned primarily with the use of the 30-channel CEPT PCM system, it is no bad thing to give detailed consideration to the several practical PCM systems in use and to the other system adopted as a standard by the CCITT for North American standard use.

The history and nature of the four systems to be discussed is, in summary:

1   **T1,** later known as **D1 system**, introduced by AT&T in 1962.
    Twenty-four-channel 8-bit-per-channel system. In each channel 7 bits are used for encoding the voice signal; the eighth bit is used for signalling for that channel, i.e. signalling at 8 kbit/sec. Frame alignment is by an extra 193rd bit per frame. *Figure 3.10*, Table 3.3.
2   **D2 system.** Standard system described by CCITT recommendation G733.
    Twenty-four-channel 8-bit-per-channel system, 193 bits per frame. Signalling by "bit stealing" in the 6th and 12th frames of a 12-frame multi-frame arrangement, i.e. signalling at 1.3 kbit/sec. *Figure 3.11*, Table 3.4.
3   **UK 24-channel system.**
    Twenty-four-channel 8-bit-per-channel system with 1 bit per channel reserved for signalling and frame and multi-frame alignment. A 4-frame multi-frame is used, i.e. signalling at 4 kbit/sec. *Figure 3.12*, Table 3.6.
4   **CEPT 30-channel system.** Standard system described by CCITT recommendation G732.
    Known as the CEPT system because the CEPT first specified it as a European standard. It is now adopted as the standard for most of the world. Thirty-two-channel 8-bit-per-channel system. Thirty channels used for speech; channel 0 reserved for frame and multi-frame alignment and for PCM link signals (error and failure messages, etc.). Channel 16 reserved for signalling and divided into two 4-bit quartets. Signalling quartets channel associated within a 16-frame multi-frame, i.e. signalling at 8 kbit/sec. *Figure 3.13*, Table 3.7.

The sections that follow describe all four systems in some detail. Understanding the various systems is an excellent way of achieving the necessary familiarity with the ideas inherent in PCM in order to render the subsequent consideration of switching techniques, if not easy, at least practicable.

The information contained in Tables 3.3, 3.4, 3.6 and 3.7 on the four practical systems is collected as a final summary in Table 3.8.

T1/D1 24-channel System

Many circuits are still carried over this, the original PCM transmission system introduced by ATT in the US in 1962 and installed quite widely in North America.

Fig. 3.10 Bell D1 24-channel PCM system

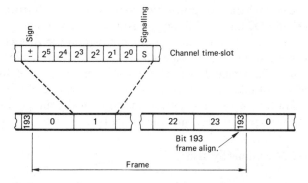

Table 3.3 Bell D1 system: summary data

Frequency range	300–3400 Hz
Sampling rate	8000 Hz
Bits per sample	7
Time-slots per frame	24
(Bits per frame)	193
PCM channels per frame	24
Output bit rate	1544 kbit/sec
Encoding law	μ-law $\mu = 100$
Signalling capacity	1 bit per channel 8 kbit/sec

Original subjective tests established that acceptable transmission of PCM encoded speech is achieved with 12-bit linear digital coding and de-coding. The transmission of at least 12 bits per channel was considered (rightly) to be excessive as it would have involved bit rates in excess of those possible over open wire lines or local trunk cable, the application originally intended for the T1 system. A 12-bit 24-channel system would require a bit rate of at least 2304 kbit/sec. This bit rate was too high to meet the criteria mentioned earlier of regenerator spacing equal to the spacing of the removed loading coils. (Experience later proved that higher bit rates were acceptable at the same regenerator spacing; see the CEPT 30-channel system described below where the bit rate is 2048 kbit/sec.)

It was at this stage that *non-linear quantisation* was proposed to reduce the bit rate and introduce a digital equivalent to FDM companding. Further subjective testing showed that companded PCM occupying 7 bits was little worse than 12-bit linear coding. These considerations, allied to a desire to use multiples of the FDM hierarchies, led to the choice of a 24-channel primary multiplex with 8 bits per channel, 7 bits to be used for speech (sign plus 6 binary digits) and the eighth bit to be used for signalling. The frame is completed by the addition of a 193rd bit to indicate frame alignment.

Signalling using an eighth bit per channel is *channel-associated* and allows simple signalling similar to loop disconnect line signalling, e.g. continuous 0000 indicating idle, continuous 1111 indicating calling, busy, answer, etc. In considering the signalling options available it is important to remember that all these digital systems are 4-wire; a system similar to that described is carrying the 24 return channels. Table 3.5 illustrates elements of the US D1 and D2 line signalling to illustrate these points.

The D1 system requires no multi-frame and is commendably simple. It is limited to 7 bits for speech and can deal with channel associated signalling only. The single 193rd bit for frame alignment means that alignment, once lost, will take time to recover. The receiving end must search through every bit in turn over many cycles, looking for the unique (long) pattern contained in the sequence of frame-alignment bits.

D2 24-channel System

Initial plans for the T1 system had envisaged its use in rural areas and for short-haul trunks. The march of events throughout the 1970s, introducing ever larger-scale integration and ever-falling costs for complex electronic circuits, continuously expanded the fields of application for digital systems. Whereas 7-bit non-linear encoding was suitable for single digital links, it was not suitable for multi-stage coast-to-coast connections where de-coding to analog and re-coding to digital might occur more than once, thus experiencing the distortions due to quantisation more than once. For similar reasons T1 was not suitable for inclusion in international connections except as a single stage. The designers of the T1 system had no intention that it should be used as anything other than a transmission system. There was almost certainly no vision of eventual digital switching and transmission. The channel associated signalling provided for T1 was inadequate for switched digital applications and there was no provision for common channel signalling (Chapter 7).

Accordingly, the D2 system was developed as an improvement on D1 and, as far as possible, compatible with D1. The design objective for D2 was to provide 8-bit non-linear coding and make possible digital common channel signalling. This could be achieved at the expense of the channel associated signalling capacity.

The D2 system is shown in *fig. 3.11* and is, like D1, a 24-channel 8-bit system with a 193rd bit provided for frame alignment. A multi-frame has been introduced consisting of 12 frames and indicated by using bit 193 of every second frame as **multi-frame alignment**. All 8 bits of all channels of frames 1–5 and 7–11 are used for voice. In frames 6 and 12, only 7 bits are provided for voice and the eighth bit is used for channel associated signalling. There is, therefore, freedom for a single signalling channel per voice channel of 1.3 kbit/sec or for two signalling channels per voice channel at 0.65 kbit/sec.

Where common channel signalling is provided, the channel associated signalling is not required and the 12-frame multi-frame is abandoned. In this case a 4-frame multi-frame is used with bit 193 of frame 2 and 4 providing a 4 kbit/sec common channel.

Pulse Code Modulation 41

Fig. 3.11 Bell D2 24-channel PCM system (CCITT rec. G733)

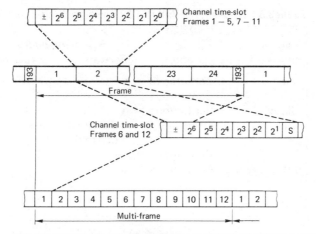

Frame no.	Frame alignment Bit 193	Multi-frame alignment Bit 193	Bit number(s) in each channel time-slot		PCM signalling channel
			for voice	for signalling	
1	1		1 – 8		
2		0	1 – 8		
3	0		1 – 8		
4		0	1 – 8		
5	1		1 – 8		
6		1	1 – 7	8	A
7	0		1 – 8		
8		1	1 – 8		
9	1		1 – 8		
10			1 – 8		
11	0		1 – 8		
12		0	1 – 7	8	B

Table 3.4 Bell D2 system: summary data

Frequency range	300–3400 Hz
Sampling rate	8000 Hz
Bits per sample	8 (normally, but 7 bits each 6th frame for channel associated signalling)
Time-slots per frame	24
(Bits per frame)	193
PCM channels per frame	24
Output bit rate	1544 kbit/sec
Encoding law	μ-law $\mu = 255$
Signalling capacity	1.3 kbit/sec

Table 3.5 Elements of Bell D1/D2 line signalling

Forward Signal	Duration	Signalling bit 8 value		Backward Signal
		Forward	Backward	
Idle	Continuous	0	—	
	Continuous	—	0	Idle
Sieze	Continuous	1	—	
Dial	10 ips	0 dial break	—	
		1 dial make		
	Continuous	—	1	Answer
Clear forward	Continuous	0	—	
		—	0	Clear back
		—	1	Blocking

The D2 system, like the D1 system, uses mu law coding (see Chapter 4) but with the value of mu now 255 (254 if the all-zero value is suppressed). Also the encoding law is adjusted slightly in frames 6 and 12 (where voice channels are still 7 bit) in order to minimise the quantisation distortion resulting from the occasional change from 8-bit to 7-bit coding.

By the time that the CCITT was ready to define a PCM multiplex there was so much 24-channel mu-law PCM installed in North America, Japan and South East Asia that the CCITT actually defined two standards. The standard met by the D2 system is described in CCITT Recommendation G733.

UK 24-channel System

The first PCM systems outside the US were introduced six years later by British Telecom (then the British Post Office) in 1968. With so much PCM already in use across the Atlantic it seemed sensible to keep the same channel capacity. There were, however, good (but occasionally questionable) reasons for altering the arrangements in more detailed ways.

The UK system was again envisaged as chiefly for the short-haul routes, primarily the local trunks known in the UK as local junctions. However, the UK uses rather more signals than are required in North America, for example meter pulses (backward), trunk offer (forward) and operator hold (backward). The eighth bit per channel, bit 1, was set aside for signalling *and* frame and multi-frame alignment leaving 7 bits for voice as in the T1/D1 system. A novel introduction was the *multi-frame* which enabled signalling capacity to be expanded and alignment to be provided without the need for a 193rd bit. The UK 24-channel system has, therefore, an output bit rate of 1536 kbit/sec.

The arrangements are shown in *fig. 3.12*, which demonstrates the availability of two 2 kbit/sec signalling channels per voice channel or a single 4 kbit/sec channel. As with the T1/D1 system common channel signalling was not envisaged* but frame 3 of the multi-frame provided an additional 2 kbit/sec channel for purposes not identified at the time.

Fig. 3.12 UK 24-channel PCM system

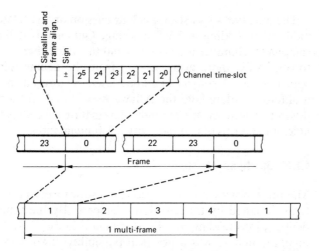

FRAME NUMBER	ALLOCATION OF BIT 1
1	Signalling for individual channels
2	Spare
3	Signalling for individual channels
4	Frame and multi-frame alignment

MULTI-FRAME ALIGNMENT PATTERN

Bit 1, channel x of frame 4	0 1 2 3 4 5 6 7 8 9 10 11 12 13 14 15 16 17 18 19 20 21 22 23
Value	1 1 0 1 0 1 0 1 0 1 0 1 0 1 0 0 0 0 0 0 0 0 0 0

Table 3.6 UK 24-channel system: summary data

Frequency range	300–3400 Hz
Sampling rate	8000 Hz
Bits per sample	7
Time-slots per frame	24
(Bits per frame)	192
PCM channels per frame	24
Output bit rate	1536 kbit/sec
Encoding law	A-law A = 87.6
Signalling capacity	4 kbit/sec

* Common Channel Signalling (CCS) has been mentioned briefly in Chapter 1. A full treatment appears in Chapter 7. At this stage the reader is asked to picture a signalling path via which processors speak to processors using messages which, if they refer at all to a voice channel, will do so by means of labels within the message. This separate signalling path could be provided on any PCM system by requisitioning one or more voice channels for the purpose. However, later PCM systems (to CCITT recommendations G732 and 733) make specific provision for CCS so as not to interfere with the voice channel modularity.

The mid 1960s had seen much discussion on the relative merits of different methods of coding and of sampling. Out of this debate emerged different sampling techniques much used in military and other systems and much valued today. Sadly, there also emerged from the debate significant differences in approach to the subject of coding laws. As a result the UK in 1968 adopted a different coding law, the A-law. As will be seen in the next Chapter, the difference between mu law and A-law in terms of coding excellence is questionable but in terms of compatibility it is fundamental.

CEPT 30-channel System

The three practical systems discussed so far were primarily transmission systems designed by transmission engineers. By the time the CEPT began its deliberations leading to a PCM standard in the early 1970s, it was clear that digital switching was a practical possibility. Two field trials of digital trunk switching exchanges had already taken place in London. The system to be standardised by CEPT was therefore the first to be considered from the point of view of both transmission and switching. By then, too, it was clear that greater bandwidths could be accommodated, albeit less easily, over the local junction network using the same repeater spacing. Control of digital switching would be more easily effected if the PCM module was based on a binary rather than a duo-decimal multiple. Combined switching and transmission would also require signalling means adequate for at least inter-register signalling and ought to be able to accommodate common channel signalling.

For all these reasons the CEPT chose a 32-channel primary multiplex: 30 speech channels and one channel reserved for frame alignment and system information and a second channel reserved for signalling (*fig. 3.13*).

For channel associated signalling, a 16-frame multi-frame is defined by the frame-alignment signal contained in channel 0 of even frames and the multi-frame alignment signal contained in channel 16 of frame 0. Channel 0s of odd frames are reserved for PCM link status indications and the channel 16s of frames 1 to 15 each carry two signalling quartets relating to two specific voice channels. Thus the signalling capacity for channel associated signalling is 2 kbit/sec.

For common channel signalling, channel 16 can be readily used as a 64 kbit/sec common channel and the multi-frame can be reduced to two frames. A further advantage of the system is that, with 8 bits per frame for frame alignment and 8 bits per 16 frames for multi-frame alignment, alignment can be regained much more quickly than with the 24-channel systems.

The coding law adopted was the A-law, retaining some measure of compatibility with the existing European PCM systems (the only ones existing at that time were the UK 24-channel systems) but finally removing all hope of compatibility with the North American systems. When the CCITT at last produced recommendations for PCM primary multiplexes, this divide was recognised internationally by defining two systems, the D2 system, Recommendation G733, and the CEPT system, Recommendation G732. Both coding laws are defined in Recommendation G711. Recommendation G711 requires

Pulse Code Modulation

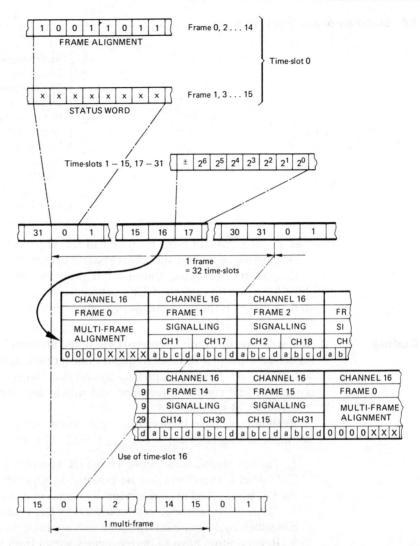

Fig. 3.13 CEPT 30-channel PCM system (CCITT rec. G732)

Table 3.7 CEPT 30-channel system: summary data

Frequency range	300–3400 Hz
Sampling rate	8000 Hz
Bits per sample	8
Time-slots per frame	32
(Bits per frame)	256
PCM channels per frame	30
Output bit rate	2048 kbit/sec
Encoding law	A-law A = 87.6
Signalling capacity:	
channel-associated	2 kbit/sec
common-channel	64 kbit/sec

Table 3.8 Summary of practical systems

> Audio-frequency band
> Sampling rate
> Bits per sample
> Time-slots per frame
> PCM channels per frame
> Output bit rate
> Encoding law
> Signalling capacity:
>   channel-associated
>   common-channel
> CCITT Recommendation

that all international links be coded to the requirements of the A-law and G732 in situations where dissimilar systems exist in the countries so linked.

Despite the increased transmission problems of a PCM system with a 2.048 Mbit/sec bit rate, the CEPT system has been received with favour internationally and is the predominant system worldwide.

Line Coding

In discussing the example system it was noted that sending binary coded signals to line resulted in a variable frequency down to DC at times, a variable power spectrum, and difficulty in picking up and maintaining synchronism. A better system of line coding is required and was in fact introduced with the first practical PCM system.

In summary, pure binary bit streams would be unsuitable as a line code for the following reasons:

1. The line signals would have a large DC component.
2. It would, therefore, not be possible to separate the line signals from the DC power feed to regenerators by means of transformer bridges.
3. The signal power spectrum would have a large component at low frequencies which could cause interference with audio circuits in the same cable.
4. Regenerators have to derive a clock signal from the bit stream and this would prove difficult with variable frequency, particularly with long series of zeros.

The process of achieving a **practical line code** is illustrated in *fig. 3.14* which uses the format of the example 4-channel system. Considerable improvement is achieved just by converting the waveform into symmetrical telegraph form where 1s (marks) are positive-going pulses and 0s (spaces) are equal-magnitude negative-going pulses. The variable frequency problem remains to a reduced extent however. The next stage is to reverse every alternate digit. Bits 1 and 3 of each channel are unaltered but bits 2 and 4 are reversed in sign. This breaks up the long string of zeros but there remains the chance that other combinations of digits will be converted into a string of zeros. For the next stage, alternate mark inversion, the clock frequency is doubled and the signal forced to cross the zero once per bit cycle. Also every alternate

Bell D1	Bell D2	UK 24-ch.	CEPT 30-ch.
300–3400 Hz	300–3400 Hz	300–3400 Hz	300–3400 Hz
8000 Hz	8000 Hz	8000 Hz	8000 Hz
7	8 (17% are 7)	7	8
24	24	24	32
24	24	24	30
1544 kbit/sec	1544 kbit/sec	1536 kbit/sec	2048 kbit/sec
$\mu = 100$	$\mu = 255$	$A = 87.6$	$A = 87.6$
8 kbit/sec	1.3 kbit/sec	4 kbit/sec	2 kbit/sec
—	4 kbit/sec	—	64 kbit/sec
—	G733	—	G732

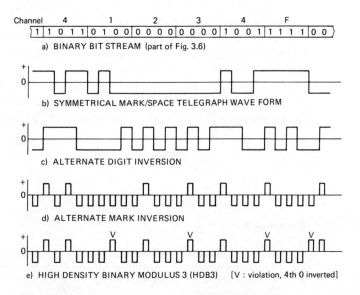

Fig. 3.14 PCM line coding

mark of the untreated signal from the alternate digit inversion stage (c) is inverted. This process has ensured regular zero crossings, thus enabling regenerators and the receiver to detect the clock frequency. Lastly, the remaining possibility of long strings of zeros is removed by inverting every fourth consecutive zero. This is done to rather complex rules and results in the High Density Bipolar modulus 3, or HDB3, line code. In *fig. 3.14e* the fourth zeros which have been inverted are marked by a V; they are known as "violations" since they result in two marks of the same polarity being received consecutively, an event precluded by the rules of alternate mark inversion.

(Try converting *fig. 3.14e* back to the binary bit stream; it will be instructive both in understanding and in appreciating the capability for complex processing inherent in modern LSI circuits.)

A further advantage of the HDB3 (or other) coding protocol is that it introduces an element of predictability into the bit stream. Regenerators and receivers can check on the validity of the bit stream they are processing by ensuring that it does indeed obey the protocol.

In discussing line coding we have strayed outside the realms of digital communications switching. Within the exchange it is quite satisfactory to use the untreated binary bit stream across the relatively short distances involved with no risk of crosstalk with audio circuits and no need to transmit timing information. The subject is necessary, however, to complete the treatment of digital transmission and the coding process and to deal with all the problems raised in considering the example.

Chapter Summary

This chapter, crucial to the understanding of digital communications switching, has provided an understanding of the process of converting analog voice signals into digital bit streams, of transmitting these digital streams, and of reconstituting the analog signals at the receiving end. An Example System with its limitations served as an introduction to the account of four practical systems, which have seen widespread commercial use since the introduction of PCM transmission and, later, switching.

Two topics remain to be dealt with in the next chapter to complete the preparatory sections of this book. These are the practical encoding laws and the synchronisation of a switched network of PCM links.

Exercises 3

3.1 How many quantising levels are available if a 6-bit binary code is used?
3.2 Explain the necessity and function of a frame-alignment signal.
3.3 Continue the series of diagrams of *figs 3.1–3.7* to show the reception process for the signal on Channel 2.
3.4 Assume that the example system is modified to provide an additional channel for signalling in the frames not carrying the frame-alignment signal (*fig. 3.15*). Prepare a new version of Table 3.2 for this modified example system.

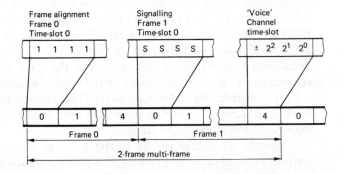

Fig. 3.15 Example system modified for signalling

3.5 A 12-channel PCM system is required to transmit electric power system telemetry signals with a frequency range of 100 Hz. What sampling frequency is required? If the coding law uses 64 quantising levels, what will be the output bit rate?

3.6 Since existing AC telephone signalling systems will be transmitted like voice over a PCM system, why do you think it is necessary to provide special signalling methods over PCM links?

3.7 List the advantages of the T1/D1 system.

3.8 List the reasons for changing to the D2 system.

3.9 Outline the differences between the UK 24-channel system and the Bell D2 system. Give reasons for the different approach.

3.10 Use you knowledge of probabilities to estimate the time required to regain frame alignment once it has been lost in
  a) The Bell D2 system.
  b) The CEPT 30-channel system.

3.11 List the differences, and the reason for the differences, between the UK 24-channel system and the CEPT 30-channel system.

3.12 In order to interpret an incoming bit stream, the receiving end has to synchronise itself with the transmitting end. What indications from the bit stream does it use for this purpose?

3.13 What minimum clock frequency is required for the CEPT 30-channel system?

3.14 As suggested in the final section, draw a diagram like *fig. 3.14* starting from the *3.14e* waveform and converting line by line back to the bit stream. Think of yourself as an LSI circuit while you are doing this; every logical inference must be based on information available, not upon your prior knowledge of the answer!

4 Error Sources and Prevention

In moving from analog to digital transmission of voice signals we have eliminated or reduced one set of problems (noise, crosstalk, distortion, etc.) but, inevitably, introduced a new set of problems. Chief among these new problems are the noise introduced by the coding process, quantisation noise, and the potential for error due to timing difficulties. The new problems are different in kind in that quantisation noise, for example, is introduced as a constant factor; it does not change with distance, transmission media, etc. Most digital error sources are of this constant nature; coded bit streams have reasonably uniform envelopes, cause known crosstalk effects on neighbouring analog and digital circuits, degenerate as a constant function of transmission distance from source, and regenerate to signals as good as source, albeit delayed in time. This chapter deals with the methods used to minimise errors and deals largely with quantisation and timing, problems which are both absent from analog transmission systems.

Quantisation

We considered the source of quantisation distortion in Chapter 3. Any system which allots a value to an analog sample falling randomly within a measurement band will produce an error proportional to the ratio of the size of the band to the typical size of the sample. As with analog transmission we will discuss **noise** as a ratio, the *ratio of the signal to quantising noise*.

Uniform Coding

With a uniform coding law, the result of the coding process and the consequent quantisation error is as illustrated in *fig. 4.1*. The signal to quantisation noise of such a linear coder can be calculated and this is illustrated for a range of linear coders in *fig. 4.2*.

Voice signals to be coded have a dynamic range of 30 dB and *fig. 4.2* shows that 8-bit uniform coding giving a signal to quantising noise ratio of about 25 dB is not quite good enough with just one coding process, let alone the many that may be required in practice.

Figure 4.1 illustrated a coding law which crossed the origin "mid-riser", that is there is no value defined for zero. To define a value for zero there must be a quantisation interval that straddles the origin "mid-tread". *Figure 4.3* illustrates the treatment of small signals (idle channel noise) by mid-riser and mid-tread quantisation. A *mid-tread quantisation law* having a quantisation interval which straddles the origin will have an odd number of quantisation intervals in total. Apart from its virtue in eliminating low level idle channel noise, a mid-tread law is also easier to manipulate using binary arithmetic because a zero is defined.

Error Sources and Prevention 51

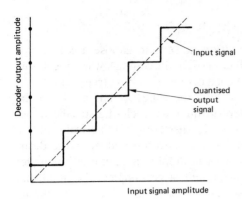
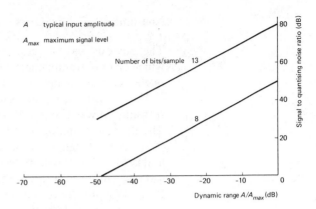

Fig. 4.2 Signal to quantising noise for uniform PCM coding of sine wave input

Fig. 4.1 Linear quantisation

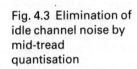
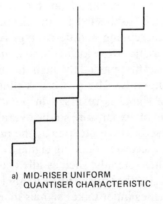
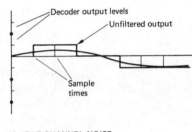

Fig. 4.3 Elimination of idle channel noise by mid-tread quantisation

a) MID-RISER UNIFORM QUANTISER CHARACTERISTIC

b) IDLE CHANNEL NOISE PRODUCED BY MID-RISER QUANTISATION

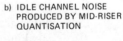
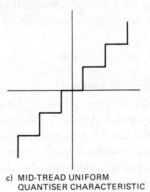
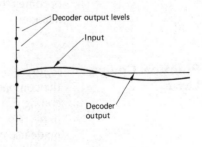

c) MID-TREAD UNIFORM QUANTISER CHARACTERISTIC

d) EFFECT OF MID-TREAD CHARACTERISTIC ON LOW SIGNALS

Non-uniform Coding Laws

The obvious way of reducing quantisation noise is to choose a coding law which gives optimum performance. Once chosen, the coding law will inevitably introduce a fixed quantisation noise at each coding/decoding transition.

The previous chapter demonstrated how the range of amplitudes to be transmitted was broken down into steps or quantas, each of a specific value. The process is analogous to that involved in answering a numerical problem "correct to n significant figures". The chapter also touched upon the desirability of coding the more frequently occurring amplitude samples, the smaller samples, more accurately than the infrequent large samples. Provided that the coding procedure is exactly reversed at the receiving end, thus eliminating this intentional distortion, then a more accurate version of the signal will be constructed at the receiver. This is the process described as companding in Chapter 2.

We can state the requirement a third way by considering the characteristic shape of *fig. 4.2* that we would have preferred. This has been added in *fig. 4.4*. Whatever signal to quantising noise ratio we achieve, we would prefer it to be constant over the full dynamic range.

Early PCM systems were developed and introduced at a time when digital circuitry was in its infancy. The transmission engineers developing these systems had no reason to change from the analog circuit methods with which they were familiar. Thus the first T1 speech coders employed a diode characteristic to perform the coding and de-coding. A matched pair of diodes having identical characteristics were located one in the coder and one in the decoder. In T1 the signal was first passed through the diode compressor and then coded using uniform intervals. The process is illustrated in *fig. 4.5*. We see two effects. First, the signal is reversed in polarity by the compression process but reversed again on expansion, so the overall effect is nil. More importantly, because the characteristic is steep near the origin, the small signals are transmitted with more accuracy than big signals, the effect which is wanted. Note that small, unwanted signals, such as idle channel noise, are also transmitted more accurately. If, therefore, idle channel noise, or background noise of any kind, approaches quiet talker signals in amplitude, then companding will make matters worse.

At the receiving end the process is reversed; waveform (iii) in *fig. 4.5* becomes the input signal and is transformed into waveform (i) by the receiver decoder.

Practical Coding Laws

Figure 4.5 illustrated an idealised coding law and assumed a system where the companding was performed first on the analog signal followed by uniform coding of the compressed signal. A block diagram of such a system is shown in *fig. 4.6*. Although the T1 system operated in this way, subsequent systems made use of digital electronic circuitry to perform the compression and expansion as part of the digital manipulation of the coding process. Practical coding laws have to meet the requirements of both minimum distortion and ease

Error Sources and Prevention 53

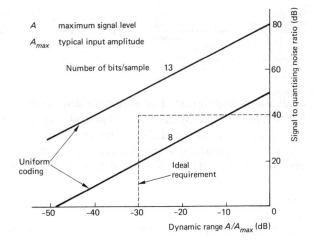

Fig. 4.4 Target performance required from non-uniform coding law

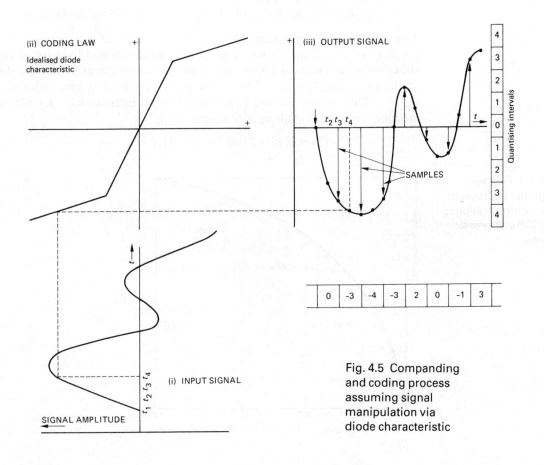

Fig. 4.5 Companding and coding process assuming signal manipulation via diode characteristic

Fig. 4.6 Companded PCM with analog compression and expansion

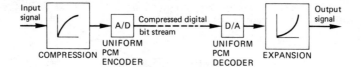

of digital manipulation. For minimum distortion, a logarithmic companding law would be ideal. Such a law will provide equal ratios of signal to distortion at all levels of input signal. For digital processing, the logarithmic companding law will be approximated by joining equidistant points on the curve by straight lines. Each straight line portion will be called a **segment**.

The μ-law

The North American systems employ a law called the μ-**law**. This law relates the output signal y to the input signal x by the formula:

$$y = f(x) = \text{sgn}(x) \left[\frac{\log_e(1 + \mu |x|)}{\log_e(1 + \mu)} \right]$$

where $\text{sgn}(x)$ is the polarity of x
and μ was defined as 100 for the T1 system and as 255 for the D2 system.

The choice of a companding characteristic with $\mu = 255$ was dictated by the property of this characteristic to be closely approximated by a set of eight straight-line segments. Further, the slope of each segment is half the slope of the preceding segment. This is illustrated for the first four segments in *fig. 4.7*. This means that the larger quantisation intervals have lengths that are binary multiples of all smaller intervals, e.g.

$$223 - 95 = 128 = 2(95 - 31) = 2 \times 2 \times (31 + 1)$$

Fig. 4.7 First four segments of straight-line approximation to $\mu = 255$ compression curve

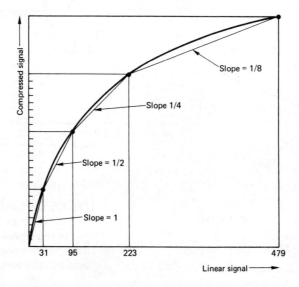

Error Sources and Prevention

Table 4.1 Encoding/decoding table for µ255 PCM

Input amplitude range	Step size	Segment code S	Quantisation code Q	Code value	Decoder amplitude
0–1	1		0000	0	0
1–3			0001	1	2
3–5	2	000	0010	2	4
⋮			⋮	⋮	⋮
29–31			1111	15	30
31–35			0000	16	33
⋮	4	001	⋮	⋮	⋮
91–95			1111	31	93
95–103			0000	32	99
⋮	8	010	⋮	⋮	⋮
215–223			1111	47	219
223–239			0000	48	231
⋮	16	011	⋮	⋮	⋮
463–479			1111	63	471
479–511			0000	64	495
⋮	32	100	⋮	⋮	⋮
959–991			1111	79	975
991–1055			0000	80	1023
⋮	64	101	⋮	⋮	⋮
1951–2015			1111	95	1983
2015–2143			0000	96	2079
⋮	128	110	⋮	⋮	⋮
3935–4063			1111	111	3999
4063–4319			0000	112	4191
⋮	256	111	⋮	⋮	⋮
7903–8159			1111	127	8031

This table displays magnitude encoding only. Polarity bits are assigned as 0 for positive and 1 for negative. In transmission all bits are inverted.
(For full definition see Table 2/G711.)

Because of this property, conversion to and from compressed code to linear representation can be performed numerically with ease.

The ingenuity put into devising such an easily manipulated rule has been overtaken by events since modern designs use table look-up methods of conversion because ROM-type solutions are now the most cost-effective.

Table 4.1 summarises the full definition table in CCITT recommendation G711 for the µ-law. Examining Table 4.1 it can be seen that each major segment of the linear approximation is divided into equally sized quantisation intervals. This is illustrated in *fig. 4.7*. The number of intervals per segment is generally 16 (strictly, there are only 15 intervals in the first segment plus the zero value). Thus an 8-bit μ-255 word is composed of: 1 polarity bit, 3 bits identifying a major segment, and 4 bits identifying a quantising interval within that segment.

This straight-line approximation to the μ-255 curve is known as the 15-segment approximation. This is because the first segments each side of zero are colinear and can be considered as one segment having 31 quantisation intervals, one interval straddling the origin (mid-tread). There are therefore two zero values: 0000 0000 and 1000 0000.

Performance of uniform μ-255 and A-law PCM encoding are contrasted later in discussing *fig. 4.8*. In brief, a μ-255 system provides a theoretical signal to quantisation noise ratio of more than 30 dB across a dynamic range of 48 dB. *Figure 4.2* reveals that uniform encoding requires 13 bits for the equivalent performance.

Table 4.2 Segmented A-law encoding/decoding table

Input amplitude range	Step size	Segment code S	Quantisation code Q	Code value	Decoder amplitude
0–2			0000	0	1
2–4		000	0001	1	3
⋮			⋮	⋮	⋮
30–32	2		1111	15	31
32–34			0000	16	33
⋮		001	⋮	⋮	⋮
62–64			1111	31	63
64–68			0000	32	66
⋮	4	010	⋮	⋮	⋮
124–128			1111	47	126
128–136			0000	48	132
⋮	8	011	⋮	⋮	⋮
248–256			1111	63	252
256–272			0000	64	264
⋮	16	100	⋮	⋮	⋮
496–512			1111	79	504
512–544			0000	80	528
⋮	32	101	⋮	⋮	⋮
992–1024			1111	95	1008
1024–1088			0000	96	1056
⋮	64	110	⋮	⋮	⋮
1984–2048			1111	111	2016
2048–2176			0000	112	2112
⋮	128	111	⋮	⋮	⋮
3968–4096			1111	127	4032

(For full definition see Table 1/G711.)

The A-law

The existence of two international coding standards, like the existence of PAL, NTSC and SECAM colour TV standards, does not reflect credit on the industry, but is entrenched and must be accepted and "lived with".

The principal idea motivating the choice of **A-law** is to make the central segments of the law exactly linear and not merely approximations to linear. The A-law is, therefore, defined in two expressions:

$$y = f(x) = \text{sgn}(x)\left[\frac{A|x|}{1 + \log_e A}\right] \qquad 0 \leq |x| \leq \frac{1}{A}$$

and

$$y = f(x) = \text{sgn}(x)\left[\frac{1 + \log_e A|x|}{1 + \log_e A}\right] \qquad \frac{1}{A} \leq |x| \leq 1$$

where $A = 87.6$, a number chosen to provide smooth transition from the linear to the logarithmic portions of the curve.

Again the A-law is expressed in 8 positive and 8 negative segments but the four segments nearest the origin are co-linear and the law is described as a 13-segment law. Table 4.2 summarises the defining table of CCITT recommendation G711.

Note that the A-law provides less protection against idle channel noise as the minimum step size is $2/4096$ whereas for the μ-law it is $2/8159$. Also the A-law uses a mid-riser format which, additionally, makes the numerical conversion to linear coding more difficult.

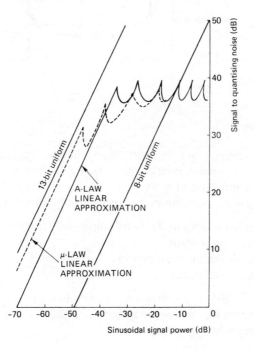

Fig. 4.8 Comparison of coding law performance

The performance of uniform coding, μ-law coding and A-law coding, is compared in *fig. 4.8*. A-law coding provides rather better dynamic range and more constant signal to noise ratio. On the debit side, the worse treatment of idle channel noise by the A-law has been noted. The difficulty of linearising A-law numerically is no longer of importance as look-up tables are now used rather than numerical manipulation. Recommendation G711 publishes the conversion tables of μ-law to A-law.

Other PCM Coding Techniques

Public telephony uses, predominantly, A-law 30-channel and μ-law 24-channel PCM encoding and the remainder of this book will assume these techniques. For completeness, however, various other coding methods are summarised here.

Adaptive Gain Control

Adaptive gain control reduces the bits required in a PCM codeword by restricting the dynamic range at the input to the coder. Automatic gain control performed this in analog systems but was not used much for telephone work because of the pointlessness of asking a talker to "speak up" if automatic gain control was operative. Adaptive gain control sends details to the receiver of the degree of gain control applied so that the receiver can apply the requisite correction (*fig. 4.9*). Used in Mobile Telephone applications.

Fig. 4.9 Adaptive gain control

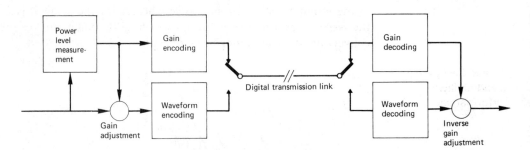

Redundancy

Voice frequency signals, particularly human speech, contain much redundancy. A two-party conversation normally (but not always!) involves only one party speaking at a time so that the duplex connection may be utilised to only 50% as a maximum. Other forms of redundancy are:

Non-uniform amplitude distribution, exploited by logarithmic coding and adaptive gain control.
Sample-to-sample correlations.
Periodic correlations.

These and other redundancies are exploited in methods to produce digitally stored speech for such applications as verbal machine response of computers, telephone exchanges, etc.

Differential Pulse Code Modulation

As its name implies DPCM methods extract and transmit only a difference indication between samples, not the whole sample. Thus redundant information is not sent but both transmitter and receiver must retain a memory of the previous sample (or samples) in order to reconstruct the signal. A particular DPCM implementation is illustrated in *fig. 4.10*. Figure *4.10a* shows a block diagram of a possible system while *fig. 4.10b* explains its operation. Both transmitter and receiver memorise the previous sample and use this sample in a feedback loop to integrate with the difference signal transmitted in order to obtain the new sample. Note that there is a possibility of a fixed error from starting conditions and, as a clock will be derived from the PCM system, the received signal will necessarily be at least one scan period behind the transmitted signal.

Fig. 4.10*a* Digital differencing DPCM implementation

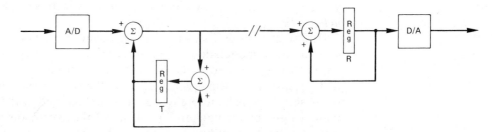

Fig. 4.10*b* Operation of differencing DPCM

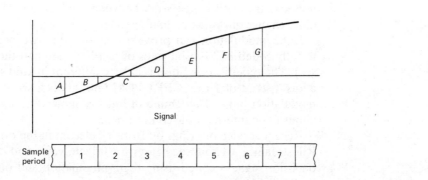

SAMPLE PERIOD	CONTENTS OF REG. T	CODEWORD TO LINE	CONTENTS OF REG. R
1	$A - y$	$A - y$	$A - y + x = A + v$
2	B	$B - (A + v)$	$B - (A + v) + v = B - A$
3	C	$C - B$	$C - B + B - A = C - A$
4	D	$D - C$	$D - A$
5	E	$E - D$	$E - A$
6	F	$F - E$	$F - A$

Initial contents of registers : $T = y$, $R = x$

DPCM methods can save 1 bit from the 8-bit sample required for full PCM. Extension of DPCM to consider, say, the last three samples could increase this saving perhaps to 2 bits, reducing necessary bandwidth from 64 kbit/sec to 48 kbit/sec.

Delta Modulation

A special case of DPCM which has been used extensively in military communications and is used in private telephony is known as Delta Modulation. In this only one bit of the difference signal is used per sample; the only signal transmitted is a 1 or a 0 specifying the polarity of the difference sample and therefore whether the signal has increased or decreased since the last sample. An approximation to the input signal is constructed in the feedback path by stepping up one quantisation level for a 1 and down one level for a 0. Thus the received signal follows the transmitted signal rather like a staircase as shown in *fig. 4.11*. A block diagram of a suitable system is illustrated in *fig. 4.12*.

DPCM introduces its own forms of quantisation noise, identified in *fig. 4.11* for delta modulation. Slope overload occurs when the signal changes too fast to be followed at the sampling frequency in use. It should be noted that slope overload will occur, if at all, at the same periodicity as the maxima of the signal and is therefore to a great extent masked by the signal. Granular noise, caused by the unfiltered remnants of the overshoots and undershoots, is continuous at sampling frequency and analogous to normal quantisation noise. A trade-off is necessary between small step sizes (low granular noise, more chance of slope overload) and large step sizes.

Delta modulation can provide qualities that are perceptually equivalent to fully coded PCM at bit rates of just less than half the PCM rates (64 kbit/sec). The quantisation noise is clearly different in kind so that CCITT, which allots a score of 1 to a CEPT PCM link, allots a score of 6 to defined delta modulation links. Tandeming of links is allowed up to a maximum score of about 14 on international connections.

Recent service offerings by British Telecom in the private network include Kilostream, a 64 kbit/sec bearer for private voice and data. A voice channel on Kilostream is coded using delta modulation and occupies 32 kbit/sec of the available capacity.

Filtration

It is appropriate now to investigate the rule that, for accurate reproduction, the pulse amplitude sampled signal must be sampled at a frequency of at least twice the bandwidth. Up to now it has been stated but not justified.

Figure 4.13 illustrates a typical pulse amplitude modulation system. The analog waveform is sampled at a constant frequency, f_s, and re-constructed using a low-pass filter which interpolates between sample values giving a smooth decoded output.

Figure 4.14 shows the process in terms of frequency spectrum. The input signal spectrum has terms from a frequency of $-BW$ to $+BW$. The PAM spectrum is that of a continuous train of impulses and therefore consists of

Error Sources and Prevention 61

Fig. 4.11 Delta modulation
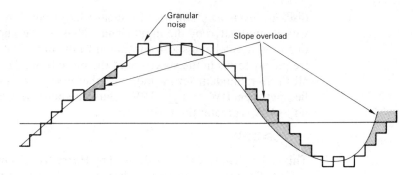

Fig. 4.12 Delta modulation system
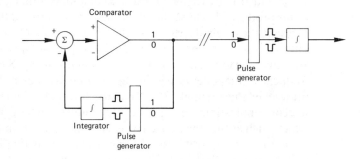

Fig. 4.13 Pulse amplitude modulation
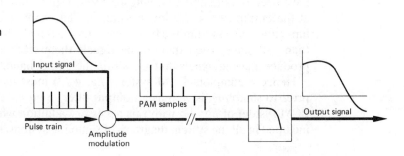

Fig. 4.14 Spectrum of PAM signal
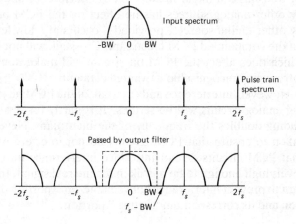

discrete terms at multiples of the sampling frequency. Each of these terms will be modulated by the input signal. Thus the output spectrum will consist of a double sideband spectrum similar to the input about each of the discrete frequency terms in the spectrum of the pulse train. The output filter removes all these redundant terms and must therefore have a cut-off frequency that lies between BW and $f_s - $ BW. Thus adequate separation is only possible if $f_s - $ BW is greater than BW, i.e.

$$f_s \geqslant 2\,\text{BW}$$

This is the *Nyquist Criterion* derived by Harry Nyquist in 1933.

Consider for a moment what happens if the Nyquist Criterion is not met. This is illustrated in *fig. 4.15*. Because of undersampling there will be distortion terms produced because the spectrum centred around the sampling frequency overlaps the original spectrum. These terms cannot be excluded by the output filter. Because of the nature of the distortion, a duplicate of the input folded back over the input, it is known as *foldover distortion*. Such distortion produces frequency components within the desired frequency band which were not present in the original. Such impairment is also known as *aliasing* both in telecommunications and in the film industry where its presence produces effects such as the stage-coach wheels rotating slowly in the wrong direction.

Figure 4.16 illustrates an aliasing process occurring in speech if a 5.5 kHz signal is sampled at 8 kHz. The sample values are identical to those that would have been obtained by sampling a 2.5 kHz signal, and this is what will appear at the output of a 4 kHz low-pass filter. This example demonstrates that the input must be band-limited before sampling to remove frequency terms greater than $f_s/2$ since, even though the signals themselves are ignored, they will produce aliasing signals that are within the output band.

Hence a complete PAM system (*fig. 4.17*) must include a low-pass filter prior to sampling as well as an output filter. *Figure 4.18* re-draws the linear PCM system of *fig. 4.6* with this necessary addition which should have been included in all the system diagrams of sampling systems shown so far.

Interference and Crosstalk

PCM transmission systems are impervious to degradation of the original signal by transmission impairments such as interference and crosstalk. Interference by other analog sources has no effect on the PCM bit stream. Interference by other digital sources, provided it occurs at a low level, will have no effect on the companded PCM bit stream. Crosstalk will not exist since interference which does affect the PCM bit stream will make nonsense of it rather than introducing recognisable unwanted elements.

However, interference and crosstalk of the PCM bit stream upon neighbouring analog circuits will be serious. It is partly for this reason that the HDB3 coding doubles the frequency of the bit stream. Nevertheless, care must be taken to ensure that PCM circuits do not interfere with analog circuits and that PCM circuits do not introduce interference into other PCM circuits at levels high enough to be mistaken for the real signal. In particular, it is important to prevent interference between the "go" portion of a PCM 4-wire connection and its corresponding "return" portion.

Fig. 4.15 Foldover spectrum caused by undersampled input

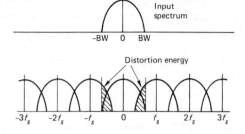

Fig. 4.16 Aliasing of a 5.5 kHz signal into a 2.5 kHz signal

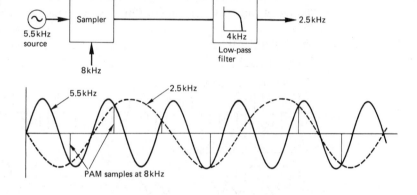

Fig. 4.17 PAM system

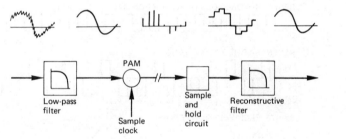

Fig. 4.18 Band-limited PCM system

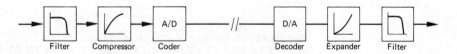

In providing PCM circuits over existing cables these considerations have to be borne in mind. Existing cables, designed for analog 2-wire transmission, consist of bundles of pairs laid up together in patterns suitable for reducing crosstalk in the voice band but not necessarily the high-frequency components of a digital bit stream. In placing PCM circuits onto a conventional analog cable, certain pairs will be unusable because separation rules (of PCM circuits, of the go and return portions of the same circuit, and between analog and PCM circuits) must be observed. The resulting "fill ratio" of circuits to available metallic pairs will be considerably less than 100%.

64 Introduction to Digital Communications Switching

Cables designed specifically for PCM circuits will have go and return paths in separate screened portions of the cable. Using conventional, existing cable the ideal is to accommodate the go and return paths in separate cables.

Timing

Two aspects of the timing problem have to be considered: *timing delay* and *timing synchronisation*. Examination of some of the diagrams introduced so far will reveal the existence of PCM timing delay. The series of figures *figs 3.1–3.7* demonstrated that, at the very least, a signal sampled at the beginning of a PCM codeword period cannot be reconstituted until the end of the period. (*Figure 3.14* and Exercise 3.14 demonstrate this more forcefully.)

As long as we consider PCM transmission only, timing delay is the only timing problem of relevance. Once we consider switching digitally between

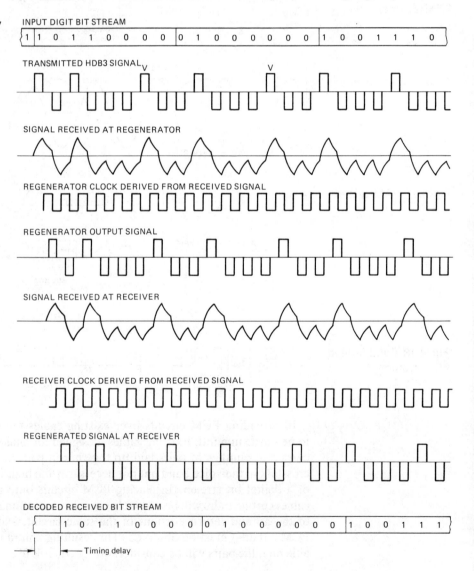

Fig. 4.19 Timing delay

different PCM systems then it is essential that the PCM systems must themselves be in synchronism. We will consider timing delay in some detail now but synchronisation will be discussed in outline only, leaving a more detailed treatment until more understanding of digital switching has been gained.

Timing Delay

The passage of a signal through PCM coding, transmission degradation, regeneration, and eventual digital-to-analog decoding, is illustrated in *fig.*

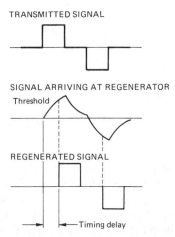

Fig. 4.20 Timing delay detail

4.19. An abbreviated but magnified version is shown in *fig. 4.20* for a single digit.

The square-wave signal sent to line is attenuated and subjected to the, mainly capacitive, impedance of the cable. It arrives at the regenerator therefore as a signal made up of segments of exponential rise and exponential fall. The receiver will have a threshold below which no signal is recognised so that there is a delay, equal to the exponential rise time to the threshold, before the presence of a signal is recognised. The receiver clock will therefore be timed to follow the transmitter clock by, at least, the cable delay constant. The regenerated signal, timed by the clock, will also be delayed. *Figure 4.19* shows these effects, with a little exaggeration, for a regenerator stage and the receiver, and indicates an overall delay of $1\frac{1}{2}$ clock periods. Any storage and processing of the digital bit stream will introduce delay over and above this minimum timing delay.

Such delays are of no great importance provided that the transmit and receive paths are subject to the same amount of delay. If, however, transmit and processing of the digital bit stream will introduce delay over and above could be a build-up of delay which would affect speech transmission and the performance of duplex data transmissions.

In analog transmission there is the delay of electrical waves through transmission lines. Introduction of digital transmission has exacerbated the effect because of the digital threshold decision values applied to the distorted received signal.

Synchronisation

Quite often in telecommunications we employ techniques such as alternate routing. This provides a good example with which to lead from the subject of timing delays into synchronisation. Suppose there were two alternative routes providing access from Exchange A (*fig. 4.21*) to Exchange B. Route 1 is direct and has 12 regenerators; route 2 is via a transit point and has 24 regenerators. Clearly, leaving other more complex considerations aside, the timing delay on route 2 will be twice that on route 1.

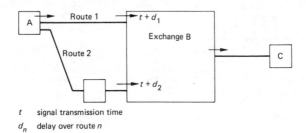

Fig. 4.21 The need for synchronisation

t signal transmission time
d_n delay over route n

Exchange B may be required to transit-switch a connection from exchange A to exchange C. If exchange B synchronises all its clocks to the PCM bit stream received on route 1, then there will be a problem in trying to switch traffic received on route 2 to exchange C. This illustration is a particular case of a general problem. If we are to switch digitally between incoming digital systems and outgoing systems, then all the incoming and outgoing systems must operate at clock rates which are identical or possess some common relationship. As the digital network spreads so the need for synchronisation spreads.

Two solutions to the problem will be discussed here: the synchronisation of a network and the use of delay to interwork between separately synchronised networks. A more detailed treatment of the subject is provided in Chapter 7.

Figure 4.22 illustrates a **network synchronisation** scheme similar to that adopted for the UK system. A *national reference clock* is used to drive the whole network via a number of secondary reference clocks which maintain mutual synchronisation amongst themselves and in turn define the local clocks at exchanges lower in the hierarchy. On failure of the national reference or the links to the second level, the secondary references maintain mutual synchronism by majority agreement among themselves. Every exchange synchronises its local clock to one of two PCM system bit streams.

The national reference will be a highly accurate clock and synchronisation of the primary level is despotic or master-slave. The primary reference level is able to revert to mutual synchronisation should the national reference fail and synchronises the secondary reference level despotically. The local exchanges are synchronised despotically by the secondary level.

Each PCM system terminating on an exchange will utilise an *aligner mechanism* such as that shown in *fig. 4.23* to delay the incoming bit stream until

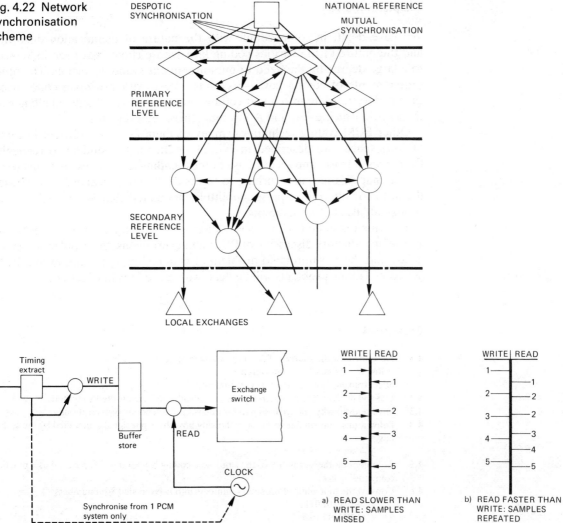

Fig. 4.22 Network synchronisation scheme

Fig. 4.23 PCM system synchronisation

Fig. 4.24 Effect of slip

it is in synchronism with the exchange clock. This mechanism will introduce the possibility of slip errors unless the network is synchronised or the exchange clocks are so accurate that slip seldom, if ever, occurs. In this latter case the network is said to be operating plesiochronously. The effects of slip are illustrated in *fig. 4.24*.

In practice, national reference clocks are made highly accurate so that the international network can work plesiochronously. Within the national network various forms of despotic and mutual synchronisation are used to reduce the occurrence of slip to acceptable levels. What constitutes an acceptable level depends on the traffic being carried by the network; slip in voice can be acceptable at much higher levels than, for example, slip in connections carrying facsimile or data.

Chapter Summary

We have dealt, in some detail, with the nature of quantisation distortion and the nature of the various coding laws whose choice has been influenced to a large degree by the need to minimise quantisation distortion. The other requirements of coding laws—maximum detail with minimum code space, the need to translate into linear coding, etc.—have been discussed along with the ways in which μ-law and A-law meet these requirements.

Other PCM coding techniques, many of them far more efficient in terms of code space, are described in enough detail for the student to recognise the names at least. The areas of use of these coding techniques are indicated.

The chapter then, in returning to topics dealt with in Chapter 2, discussed the need for filtering of pulse amplitude modulated signals, not only on decoding but also before sampling.

The chapter ended with a short discussion of interference and crosstalk and an introductory discussion of PCM timing problems, the problem of timing delay, and the not-unrelated problem of maintaining synchronism in a PCM network. Timing problems will be dealt with more fully in Chapter 7.

Exercises 4

4.1 Describe how the sources of error in analog telephony:
  loss   crosstalk   interference
affect transmission in a digital environment.

4.2 List and describe the new sources of error introduced by using digital methods.

4.3 Give reasons why idle channel noise is offensive in telephone conversations.

4.4 Draw a diagram similar to *fig. 4.5* showing how the input signal is modified by the coding law.
$$y = \sin x \quad -1 \leq x \leq 1$$

4.5 Comment on the suitability of the sinusoidal coding law used in Exercise 4.4 as a practical companding law.

4.6 Convert the following 12-bit binary numbers into μ-law coded 8-bit numbers:
  0010 0101 0111
  0010 1011 1000
  $-$0000 0010 1010
  0001 0111 0000
  $-$0110 0000 1100

4.7 Convert the 12-bit binary numbers given in Exercise 4.6 into A-law-coded 8-bit numbers.
  Note If you refer to the full tables in G711 to assist you in answering Exercises 4.6 and 4.7, award yourself extra marks for enterprise but be careful to note that the tables show transmitted codes, reversed compared with Tables 4.1 and 4.2 given in this chapter.

4.8 Determine the sequence of code words required to transmit a 1 kHz sine wave at half maximum power. Do this first for μ-law then for A-law. (This question, or part of it, is provided as a worked example in the answers, page 202.)

4.9 Convert the following A-law codes into μ-law and vice versa. This time you are advised to consult G711 but remember to convert as the codes below are written "as coded" and G711 lists codes "as sent".

  A-law codes   0,010,1001 μ-law codes   1,001,1001
                 1,100,1101   0,110,1111
                 1,111,0000   1,000,0101
                                                    0,001,0111

4.10 A private exchange is to be provided with a voice announcement system to guide the extension users in the operation of its sophisticated features. Suggest suitable coding laws for digitally recorded speech for this application.

4.11 What new sources of error are introduced by delta modulated PCM?

4.12 What will be the frequency of the aliased signal if a 6 kHz signal is sampled at 8 kHz?

4.13 PCM signal regenerators are designed to recognise signals at 40% maximum. Cable sections between regenerators have time constants such that 40% signal is reached 15 μsec after signal transmission. The cable distance between the transmitter and the first regenerator and last regenerator and receiver is half the normal span between regenerators. The equivalent time constant for these half sections can be assumed to be 9 μsec. (This is normal practice.) What is the total delay exchange A to exchange C if exchange B introduces 275 μsec and there are 7 regenerators A to B and 12 regenerators B to C?

5 Digital Exchanges and Switching

Introduction

We have at last arrived at the central theme of the book: PCM communications switching. This chapter takes as its subject matter the shaded portion of *fig. 5.1*, the switching function. It will be dealt with in two stages: first, concentrating on distribution switching, and this in a fairly theoretical manner, and then turning to concentration switching using a modern practical system as an example. This order of treatment is in line with historical development. Historically the first digital exchanges were trunk exchanges (Empress, London 1968; Moorgate, London 1971) and we are still awaiting the moving of the coder into the subscriber's instrument. Yet, as we have shown, the great advantage of PCM of combining connection switching and transmission (multiplex) switching is only fully exploited when the PCM connection extends from end to end of the network.

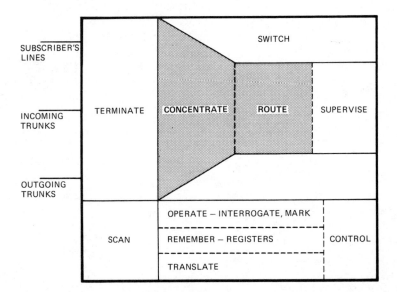

The reasons why development has not followed the path of greatest benefit are worth examining in the context of the allied developments that were occurring in the same decades. The previous chapters have demonstrated clearly the considerable technical complexity involved in converting analog signals into PCM bit streams. The equipment required to perform the conversion and re-conversion was at first constructed from discrete components: transistors, diodes, resistors, capacitors, etc. It was big, expensive and relatively greedy in power consumption. As a result administrations could only afford to introduce such equipment at the centre of the network, operating on traffic that was already concentrated. Gradually the introduction of successive stages of integration (many components provided on a single chip of silicon) reduced the bulk, the cost and the power requirement.

At the periphery of the network the exchange interface with the subscriber's line and telephone instrument had, and often still has, to provide line feeding currents to carbon microphones, high-energy currents to sound the telephone bell, and, perhaps, recognition of loop disconnect dial impulses. All these are functions largely incompatible with low-powered integrated devices.

Fig. 5.2 Cost distribution of exchange functions

	(a) ANALOG CROSSBAR	(b) DIGITAL SPC
TERMINATE	10	70
SCAN	5	5
SWITCH	50	5
CONTROL	35	20
	100	100

Effectively, the trend to move the codec (coder/decoder) stages of the digital network out towards the subscriber also moves the cost of the network in the same direction. If the functions of *fig. 5.1* are realised in analog technology, they will have a cost distribution as shown in *fig. 5.2a*. Figure 5.2b, by contrast, shows the equivalent cost distribution for a digital exchange. The figures of *fig. 5.2* are very tentative; proportions change with the size of the exchange. However, the trend is very clear. The resulting problems of designing a cost effective line interface including codec functions, either as an exchange line interface or, eventually, in the telephone itself, are massive and are even now only approaching satisfactory solutions.

The local network, the telephone and the local line plant exist and represent too large an investment for rapid replacement so that all viable digital solutions to extending the digital network towards the subscriber must operate with existing instruments and line plant. One immediately obvious problem is that the local line plant is two-wire whereas the digital network is essentially four-wire (see Chapter 7).

Switching: Analog Techniques and Blocking

Before dealing with digital switching it will be helpful to extend the brief discussion of Chapter 1 and study some of the principles of analog space-division switching. In Chapter 1 we introduced co-ordinate switching arrays and considered how the number of crosspoints required to connect a given number of inlets and outlets could be dramatically reduced by sub-dividing one big co-ordinate switch into two stages made up of several smaller switches.

Chapter 1 did not specify the essential requirement that inlets to a switching structure are, in general, telephone lines and that the outlets from the structure are to be connected to the same telephone lines to provide a fully interconnected network (*fig. 5.3*). Such a fully interconnected matrix is shown in two

72 Introduction to Digital Communications Switching

Fig. 5.3 Fully interconnectible switch

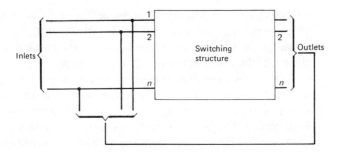

Fig. 5.4 Switching matrices

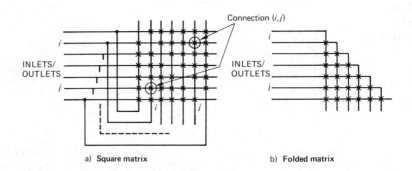

a) Square matrix   b) Folded matrix

versions in *fig. 5.4*. Note that the square matrix has, not N^2 crosspoints, but $N(N-1)$ crosspoints because there is no requirement to connect each inlet to itself. Note also, that there are two possible connections of each inlet: (i,j) or (j,i). Hence, a saving in crosspoints can be effected by using the folded array shown in *fig. 5.4b* where there is only one possible connection between each pair of inlets. This saving will be at the expense of control complexity. Alternatively, if the square matrix is retained, then an essentially two-wire matrix can be used for four-wire connection by utilising both the crosspoint options at once.

Suppose we advance further than the suggestions of Chapter 1 and consider a switching array consisting of three stages. A generalised form of such an array is shown in *fig. 5.5*. The links, sometimes called junctors, between the ranks of switches are distributed in exactly the same way as was shown in Chapter 1, *fig. 1.7*. Note that this arrangement defines to an extent the sizes of the switches. If k is the number of outlets from each first stage switch, then there must be k switches in the second stage. If inlets and outlets are equal, then the second stage switches will be square, having as many inlets (and outlets) as there are switches in the first (and third) stages.

The total number of crosspoints in such an array will be

In the first and third stages $\dfrac{N}{n} \times nk = Nk$ each

In the second stage $k \times \dfrac{N^2}{n^2}$

$$\text{Total crosspoints } N_T = 2Nk + k\frac{N^2}{n^2} \tag{5.1}$$

where N = number of inlets and outlets
n = number of inlets per first-stage switch
  = number of outlets per final-stage switch
k = number of second-stage switches.

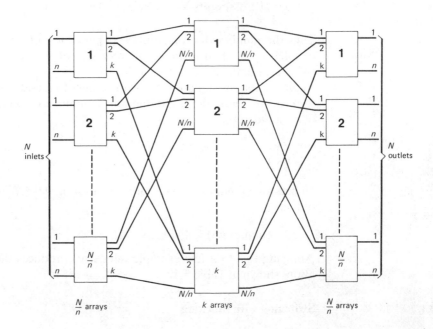

Fig. 5.5 Three-stage switching structure

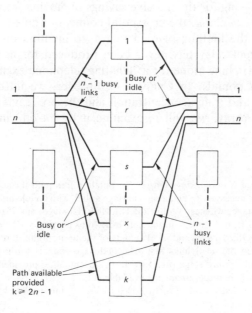

Fig. 5.6 Non-blocking three-stage network: worst-case condition

For such an array to be non-blocking, then, provided each individual array is non-blocking (i.e. all crosspoints are fitted), the number of second-stage switches k must be at least equal to $2n - 1$. This result, attributed to Clos, is explained in the footnote† and *fig. 5.6*.

Substituting for k in equation 5.1 shows that for a non-blocking 3-stage switching network.

$$\text{Total crosspoints } N_T = 2N(2n - 1) + (2n - 1) \frac{N^2}{n^2} \tag{5.2}$$

Thus the number of crosspoints is dependent, not only on the total number of inlets, but on the choice of inlet size of the first-stage and third-stage switches, i.e. the choice of the value of n.

Differentiating equation 5.2 with respect to n and equating the differential to zero, we obtain the result that for minimum crosspoints the optimum value of n should be

$$\left(\frac{N}{2}\right)^{\frac{1}{2}}$$

Hence for minimum crosspoints, substituting $n = (N/2)^{\frac{1}{2}}$ in equation 5.2 we obtain

$$N_T(\text{minimum}) = 4N(\sqrt{2N} - 1) \tag{5.3}$$

Applying this to a few example inlet sizes produces the minimum configurations shown in Table 5.1.

Switching with Blocking

Even allowing for the massive savings in moving from a single stage to three stages of switching, the crosspoint counts of Table 5.1 are still fairly horrendous. In the lifetime of reed relays, to mention one possible crosspoint, a four-contact relay has never been produced for as little as £1 so that the cost of switching alone for a 2000-line telephone exchange would exceed half a million pounds. In Chapter 1, however, we introduced the concepts of blocking and traffic concentration and, before leaving the three-stage space-switched array, we will re-examine it by introducing appropriate amounts of blocking.

---

† *Three-stage Non-blocking Arrays* To obtain a free path through a three-stage array such as *fig. 5.5*, it is necessary to find a second stage switch with an idle link to the appropriate first-stage and third-stage switches. With n inlets to the first stage, and if one inlet requires connection, then, at worst, $n - 1$ inlets are already busy, thus busying $n - 1$ links to the second stage, $n - 1$ links to the third stage, and $n - 1$ outlets of the appropriate third-stage switch. The worst case will be when the $n - 1$ links first-to-second-stage connect to different switches than the $n - 1$ links second to third stage (*fig. 5.6*). In this eventuality, provided another second-stage switch exists, then there must remain one available path.

Thus, provided $k = (n - 1) + (n - 1) + 1 = 2n - 1$, there will be a path available.

Table 5.1 Switch sizes: non-blocking networks

Number of inlets	Number of crosspoints in non-blocking network	
	Minimum 3 stage	Single stage
32	896	992
128	7 680	16 256
512	63 488	261 632
2 048	516 096	4 192 256
8 192	4 161 536	67 100 672
32 768	33 423 360	1 074 M
131 072	268 M	17 179 M

Table 5.2 Three-stage switch designs for blocking probability of 0.002 and inlet traffic of 0.1 E per inlet

Number of inlets	n	k	B	Number of crosspoints	
				Blocking	Non-blocking
32	8	6	0.75	480	896
128	8	5	0.625	2 560	7 680
512			0.4375	14 336	63 488
2 048			0.3125	81 920	516 096
8 192	64	15	0.234	491 520	4.2 M
32 768	128	24	0.188	3.1 M	33 M
131 072	256	41	0.16	21.5 M	268 M

Without developing the expression, the blocking B of a three-stage array is

$$B = \left[1 - \left(1 - \frac{p}{B}\right)^2\right]^k \tag{5.4}$$

where p is the probability that an inlet is busy, and

$$B = \frac{k}{n}$$

k = number of second-stage switches
n = number of inlets per first-stage switch
 = number of outlets per final-stage switch.

From this relationship, the crosspoint count of the networks of Table 5.1 are shown for a blocking network in Table 5.2. (Some columns of Table 5.2 are left intentionally blank to form part of the chapter exercises.)

76 Introduction to Digital Communications Switching

Table 5.3 Comparison of switching networks in terms of crosspoint efficiency

Number of inlets	Three-stage blocking network	
	Crosspoints	XPTS/inlet
32	480	15
128	2 560	20
512	14 336	28
2 048	81 920	40
8 192	491 520	60
32 768	3.1 M	96
131 072	21.5 M	164

XPTS is short for "crosspoints"

Table 5.3 provides a final comparison of the three examples discussed comparing them in terms of crosspoint efficiency. These networks have been very theoretical examples. A practical network, the TXE4 SPC system using reed relay crosspoints requires 23 crosspoints per subscriber in a local urban exchange application dimensioned to BT's normal rules. Note that the figure is 23 per subscriber. In an urban exchange, in addition to the terminations for subscribers, there will be terminations for all the trunks to neighbouring exchanges and to the national and international networks. The average figure of 23 per subscriber includes the crosspoints for these purposes also.

Switching: the Digital Requirement

The requirements of space division switching can be condensed into a diagram such as *fig. 5.7*. A similar diagram, *fig. 5.8*, attempts to encapsulate the requirements of time division switching.

The switching network is now dealing with systems, each carrying as many conversations as there are time-slots in the frame. The requirement is to switch a particular channel of a particular inlet system to a particular channel of a particular outlet system. In the case of *fig. 5.8*, channel 3 of system M is switched to channel 11 of system Y. The switch is therefore both in space (system M to system Y) and time (time-slot 3 to time-slot 11). The provision for both-way conversation is met by also switching channel 11 of system N to channel 3 of system X to carry speech in the reverse direction. The dotted additions to *fig. 5.8* demonstrate the need to associate the two-wire telephone, via 2-wire to 4-wire conversion, with system M time-slot 3 for transmit and system X time-slot 3 for receive. This part of the time division switching requirement will be discussed in connection with subscribers' line equipment and switching concentration.

We started this chapter by demonstrating the dramatic savings obtained by sharing crosspoints between potential individual conversations. In space division switching, however, the same crosspoints are used throughout each conversation. Moving to time division switching, we are discussing systems where the connection is established and dismantled separately and repeatedly for each sample of the conversation.

Three-stage non-blocking network		Single-stage non-blocking network	
Cross-points	XPTS/inlet	Cross-points	XPTS/inlet
896	28	992	31
7 680	60	16 256	127
63 488	124	261 632	511
516 096	252	4.19 M	2 047
4.2 M	508	67.1 M	8 191
33 M	1 020	1 074 M	32 767
268 M	2 044	17 179 M	131 071

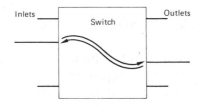

Fig. 5.7 Space division switching: the requirement

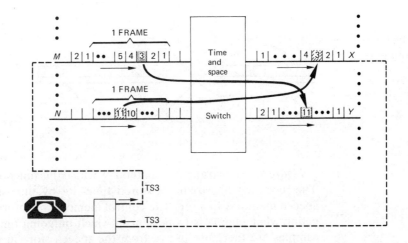

Fig. 5.8 Space and time division switching: the requirement

Returning to *fig. 5.8* the requirement shown there pictorially can be stated in words:

> Time division switching is accomplished by connecting a particular time-slot of a particular inlet system to a stated time-slot of a stated outlet system for the duration of the sample time.

Maintenance of this particular connection relationship must leave free the possibility of freely connecting any other inlet system and time-slot to an arbitrarily chosen outlet system and time-slot within the total capacity of the switch. Note that we are describing a process of switching in space (system to system) and in time (channel to channel). Note also that time division switching only is being discussed; the requirement does not demand PCM TDM.

The switching in time from time-slot to time-slot implies that the sample must be stored. In *fig. 5.8*, channel 3 was stored at time-slot 3 time to be read out at time-slot 11 time. Similarly, on the return path, channel 11 was stored to be read out on the succeeding frame time-slot 3. This need favours the use of PCM TDM because of the ease of storing digital code rather than analog samples. Analog samples are even more subject to attenuation and distortion than are the full analog waveforms. A feature of a digital switching system is that the binary 1s and 0s are regenerated at each logic stage in the system.

We have identified the two elements of the TDM switch: to switch in space and to switch in time. Each will be considered in turn before returning to discuss a practical realisation of the requirements of *fig. 5.8*.

Time Switching

The requirement of a time switch is illustrated by *fig. 5.9*. It must be capable of transferring the contents of any particular inlet time-slot into any specified outlet time-slot and be able to perform this function for every other inlet/outlet time-slot pair in any arbitrary pairing.

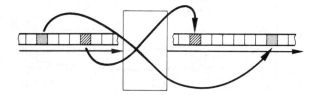

Fig. 5.9 The time switch requirement

Figure 5.10 illustrates one method by which this function may be performed. The incoming bit stream is stored time-slot by time-slot as it arrives in a speech memory. In a separate control memory the information is stored indicating what sample is to be sent in which outgoing time-slot. The outgoing samples are therefore drawn from the speech store in the order defined by the control memory. Thus, sample value x received in time-slot i is stored in time-slot numerical order in the speech memory. The outlet bit stream is made up of samples as directed by the control memory so that sample value x is output in outlet time-slot j. The system shown is capable of duplex connections if, when outlet time-slot j is allocated to inlet time-slot i, then outlet time-slot i is allocated automatically to inlet time-slot j.

If the input to and output from *fig. 5.10* is a 30-channel CEPT system, then the diagram represents a method of providing full availability switching for 32 conversations. Allowing for timing and control circuitry such an arrangement could be realised with just parts of three or four medium-scale integrated

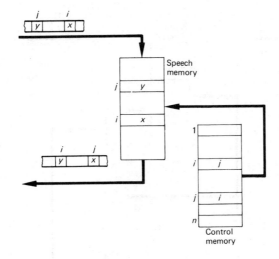

Fig. 5.10 Simple time switch time-slot interchanger

circuits. This is clearly a very great reduction on the equivalent space division switch needed to provide space division switching for 32 inlets and 32 outlets.*

The number of channels c that a time switch such as *fig. 5.10* can accommodate is given by

$$c = \frac{\text{Frame time}}{2 \times \text{Memory cycle time}} \tag{5.5}$$

For an 8 kHz sampling rate and a 500 ns memory cycle time,

$$c = \frac{125 \times 10^{-6}}{2 \times 500 \times 10^{-9}} = 125 \text{ duplex channels or 62 connections}$$

The Time Switch Stage

The previous discussion and *fig. 5.10* dealt with the time-slot interchange function only. A realistic design of time switching stage might have peripheral multiplexing equipment. Many practical time switches switch parallel PCM rather than serial PCM and this requires series-to-parallel conversion at input and the reverse at output (unless succeeding and preceding stages also switch in parallel form).

A further refinement often employed is to increase the speed of operation of the switching stages. An example might be to quadruple the speed so that each inlet and outlet of *fig. 5.10* represented four PCM systems rather than one system. This approach is often allied with parallel switching, thus reducing the necessary increase of switching speed.

---

* In switching 30-channel CEPT systems, we can use all 32 channel time-slots for conversations as there is no need to switch the framing data in transmission channel 0 nor the signalling data of channel 16. This will be explained further towards the end of this Chapter.

Fig. 5.11a Time-slot interchange: sequential Write, random Read

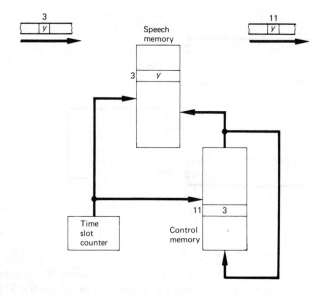

Fig. 5.11b Time-slot interchange: random Write, sequential Read

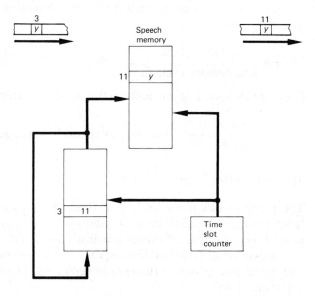

Figure 5.10 demonstrated time-slot interchange using a sequential writing process followed by a random reading process. The order of the two processes can be reversed. An example of each approach is illustrated in *fig. 5.11*. The sequential writing approach can be described as output-associated control whereas the sequential read method is input-associated control. In certain designs of time and space switch arrays it might be convenient to have time switch stages in either form.

The physical realisation of such a time switch stage will probably be achieved using one or more customised integrated circuits. Such a circuit for a time

Fig. 5.12 Time switch element (no. 4 ESS)

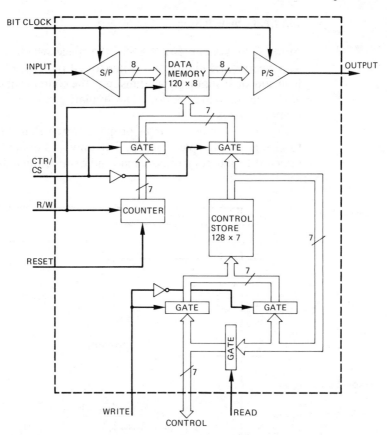

switch element is illustrated in *fig. 5.12*. This and following examples (*figs 5.12–5.14*) are drawn from the ATT No. 4 ESS trunk switching system design.

Figure 5.12 is a design implemented on a single LSI circuit. One hundred and twenty channels use a control store of 128 words, which allows some switching expansion. The configuration of output-associated or input-associated control is determined externally by a control which selects between time-slot counter for sequential accesses and control store for random accesses. The control store, an "end-around" shift register, is surrounded by circuitry enabling control information to be written in and read from the store. The control leads shown in *fig. 5.12* are defined in terms of a particular system, No. 4 ESS, and represent the requirements of that system, but, in general terms, we recognise the provision for writing new control data (WRITE), interrogating existing control data (READ), requiring sequential or random access (CTR/cs), and clocking into the data memory at bit rate rather than PCM word rate (BIT CLOCK).

The Space Switch Stage

In the same system (No. 4 ESS), the space switch stage is also implemented as a custom LSI circuit and is illustrated in *fig. 5.13*. One output is selected from any of 16 inputs depending upon the contents of the control store pointed at during that clock period. The similarity between this design and *fig. 5.12* is not accidental.

Figure 5.14 shows an array of space switch elements like *fig. 5.13* arranged to give a 16×16 space switch. The only additional elements are two 1×16 selectors to gate control store read and write commands to individual elements.

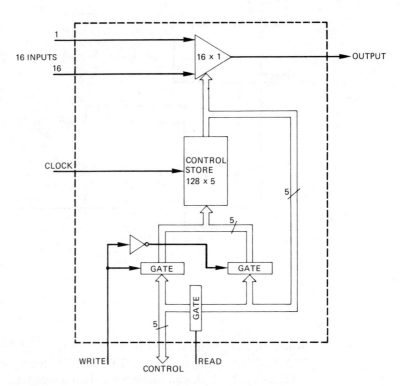

Fig. 5.13 Space switch element (no. 4 ESS)

Time and Space Switch Arrays

We are now in a position to begin assembling these elements together to produce combined arrays switching in time and space and providing the complete communications switching function. A simple arrangement of a time switching stage followed by a space switching stage is illustrated in *fig. 5.15*. Note that a time division space switch is re-configured at each time-slot; the information "holding" each conversation path is stored in the various control memories.

For full connectivity it is clearly necessary to have both time switch and space switch functions in some form. The capacity of the basic two-stage

PCM Exchanges and Switching   83

Fig. 5.14 16 × 16 time division space switch

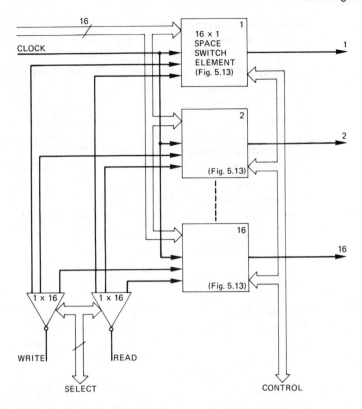

Fig. 5.15 Time/space switching configuration

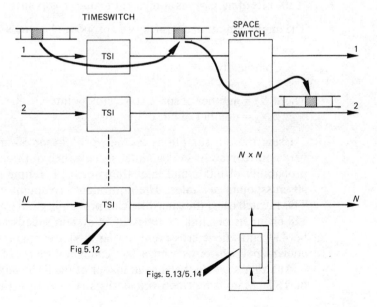

switch can be further enhanced, first by time-multiplexing at the input (and therefore having to switch at higher speed), and next by adding more stages. Three-stage arrangements have, as in space division systems, proved effective and both a time switch between two space switches (STS) (*fig. 5.16*), and a space switch between two time switches (TST) (*fig. 5.17*), have been used.

The choice between STS and TST configurations is very much influenced by the overall system design concept. The TST arrangement gives opportunity to share second-order multiplexing and serial-to-parallel conversion functions with peripheral parts of the network. In the No. 4 ESS system the central space switch was expanded into three stages (final configuration TSSST). The possibilities for "stretched" versions of original designs are more easily realised in systems using a central space switch.

An advantage of STS arrays for smaller networks is that the network can include a degree of concentration or expansion. This is not possible with a single space switch in a TST configuration.

Switching Network Comparison

The introduction of time switching and electronic crosspoints has made it difficult to compare TDM switching with the space division examples considered earlier. Any such comparison must recognise factors such as the level of integration used in the systems compared and the speed of switching used in the time switching stages.

Reference 5.1 proposes a comparison method for systems using medium-scale integration and equal switching speeds. In this comparison the following equivalencies are assumed.

One AND gate is equivalent to one crosspoint.
One AND gate is equivalent to one-and-a-half IC pins.
100 bits of memory is equivalent to one crosspoint.

The implementation complexity so proposed becomes:

$$\text{Complexity} = N_x + \frac{N_B}{100} \tag{5.6}$$

where N_x = number of space stage crosspoints
N_B = number of bits of memory.

Using these assumptions the networks so far discussed can be compared very approximately. A 2048 inlet space switch of three stages with a blocking probability of 0.002 and inlet traffic of 0.1 E requires 81 920 crosspoints or 40 crosspoints per inlet. The equivalent crosspoint count for a three-stage TDM switch using the same blocking probability and traffic loading, and having 128 channels per link (4 times PCM system speed) and 16 TDM links, will be 430 equivalent crosspoints using a STS configuration (*fig. 5.16*) or 656 equivalent crosspoints using a TST configuration (*fig. 5.17*).

At larger sizes the balance in favour of the STS configuration tips in favour of TST. Many small network designs start with a time switch and "grow"

Fig. 5.16 Three-stage STS array

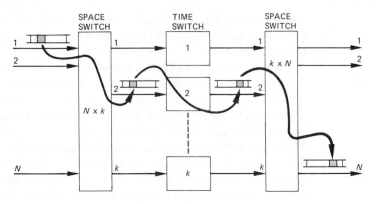

Fig. 5.17 Three-stage TST array

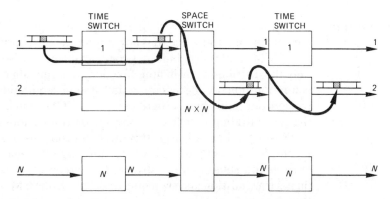

into three-stage networks by the addition of space switches, that is they start as T and grow to STS. Large network designs which retain the separation of time and space switch functions usually employ the TST structure.

Switching Delay

The time-slot interchange process of a time switch introduces a delay of a magnitude that completely eclipses the delays due to transmission of PCM discussed in Chapter 4. (The attempt to answer exercise 7 will demonstrate this very clearly.) There, delays of at least 1 bit (3.91 μsec in the CEPT system) possibly extending to say 3 bits (11.72 μsec) were considered. A single time switch will introduce an average delay of half the scan rate (62.5 μsec) variable between zero and 125 μsec. This could be a reason for choosing STS configurations, limiting the switching delay to one time switch contribution.

The problem is one which will figure repeatedly in the remainder of this book. It has little effect on fully PCM encoded voice signals but can affect data transmissions between communicating machines over the voice network. This is particularly true if the go and return paths of a duplex connection experience different delays, as they certainly will do unless the system is designed specifically to prevent this. Switching delay will also affect differentially encoded PCM as the receiving end has to retain its memory of the previous sample for a variable period. It will affect any attempt to switch more than one channel between the same destinations, a feature required for wideband circuit switching.

Early experimental PCM switching systems were designed to minimise the problem by using switching algorithms which favoured identical time-slot interchanges and progressed through a hierarchy of less-favoured connections minimising the switching delay. Such algorithms increased the size and complexity of the switch and, more importantly, greatly complicated the control. Modern system design tends to ignore the problem in the switch functions but to allow for its effects particularly on the connections most affected. One approach is to employ routing methods which constrain the return path to a switch routing which introduces the same delay as that on the go path.

Time Switching in Practice

So far the contents of this chapter have dealt largely with the theory of switching both in time and space. It is now appropriate to consider a practical example of a PCM TDM switching network. Until now, also, we have considered only the distribution switching function, ignoring almost completely the concentration function necessary in any system required to economically switch traffic from individual subscribers. Early PCM switching systems performed the concentration switching on the analog connections in space division prior to PCM encoding. Clearly, this method economises on PCM coding equipment. More recently it has become more urgent to encode at an earlier stage, even at the subscriber's instrument or premises, extending the benefits of digital transmission and switching to the subscriber. Modern switching systems therefore have at least the option to encode in the line interface circuit, and concentration switching is then performed digitally.

Switching Architecture

Before discussing a particular example it is appropriate to review the architecture of switching systems, whether space division or time division.

Figure 5.18a re-draws the switching portion of *fig. 5.1* in more detail. The switching function must concentrate the traffic from subscribers' lines and if possible distribute it so that particular subscribers are not always impeded from making calls by their association with other lines which may be high calling rate. (If your line is permanently competing for scarce, concentrated switch outlets with another line which is frequently busy then, although the exchange as a whole gives adequate service, your experience is quite the contrary.)

The distribution and routing stages can be formed of assemblies of switches similar to those we have discussed. These stages ensure that every exchange terminal and every circuit of every trunk route is available to every line and trunk termination. Note that traffic local to the exchange has to go through an expansion stage, reversing the function of concentration. In many systems the functions of concentration and expansion are performed by the same switch.

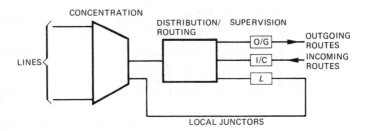

Fig. 5.18a Generalised switching architecture

Fig. 5.18b Step-by-step architecture

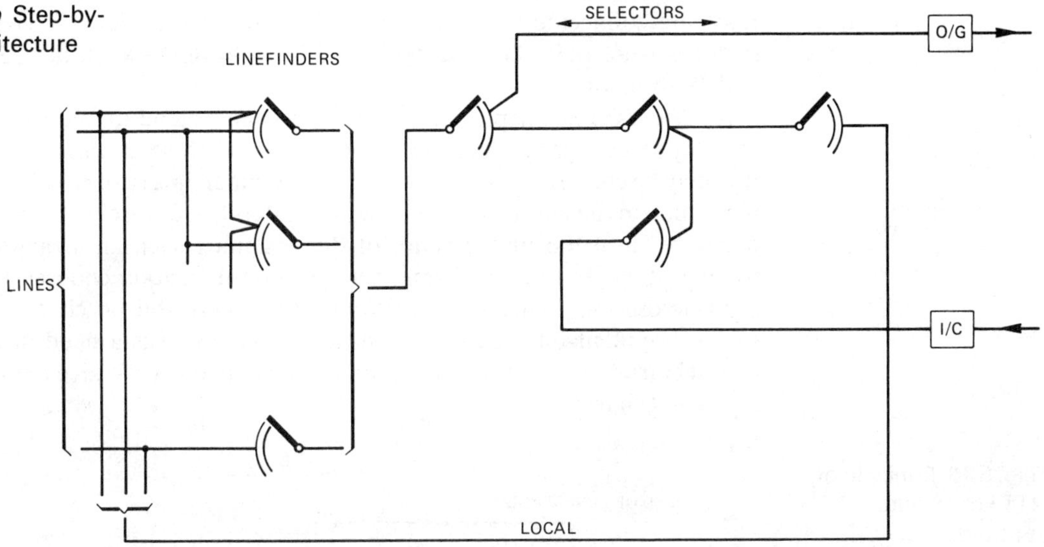

Two tendencies can be identified in the practical switch designs used in the past or in use today. Their comparative merits are defined to a large extent by the methods of control employed and, as such, are discussed later. One tendency is towards *step-by-step switching*, the other towards *link frame switching* typified perhaps by the Bell No. 5 Crossbar system.

Figure 5.18b illustrates the step-by-step approach. Concentration is provided by the line finder stage although further concentration (or expansion) is possible in subsequent stages. In heavy traffic applications the line finders are turned around to become subscribers' selectors, each subscriber having one dedicated selector in the concentration stage. Routing distribution is spread over subsequent stages; the heavy traffic routes being taken from earlier selector stages than those used for light traffic routes. Incoming routes are switched over dedicated selectors sharing the same multiple as local traffic. Impeding of calls is minimised by the high availability of the switches (25 or 50 in line finder stages typically). Note that switching equipment is saved by separating out trunk traffic midway through the switch. Conversely, step-by-step systems must provide separate expansion stages for terminating traffic; it is not normally possible to share originating and terminating concentration.

Fig. 5.18c Frame architecture

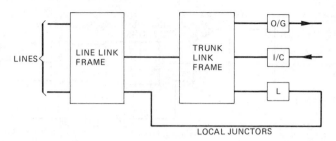

A link frame (*fig. 5.18c*) was, typically, a two-stage array of crossbar switches fully interconnected in a link arrangement like that of *fig. 1.7*. The crossbar switches were quite large (a typical ITT Pentaconta switch has 22 verticals and 28 horizontals).

As discussed in Chapter 1, step-by-step systems tend to select a free outlet at every stage regardless of the situation thereafter, so that a false busy situation may be created. (As an aside, the first commercial use of crossbar switches, in Sweden in the late 1930s, used them in a step-by-step architecture, probably partly because the mathematics of dimensioning such an arrangement are somewhat easier.) In link frame systems, a frame marker chooses a free path right across the frame but inter-frame links may still be chosen that lead to a subsequent busy path. For these reasons there is a need in all switch architectures, either to choose a path from end to end or to give the subscriber a "second chance".

Fig. 5.19 Principle of ITT Pentaconta Entraide

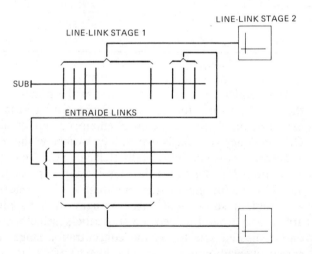

One such "second chance" arrangement is the feature known as "entraide" in the ITT Pentaconta systems. This is illustrated in *fig. 5.19*. The final outlets available to a subscriber are trunked, not to an inter-frame link but to an entraide link which has a terminal appearance on the line link frame like that of a subscriber. Thus, the subscriber meeting busy on normal switch outlets is given a second chance from a new starting point, thus increasing the chances of completing a path. In one form or another such second chance features are present in most practical systems although seldom with the elegance of the Pentaconta method.

PCM TDM Concentration

In discussing a practical system we are assuming a system where coding is performed in the line interface circuit at the individual line termination on the exchange. Such an arrangement was far too expensive to be contemplated on early PCM local exchange systems but is now becoming competitive as suitable LSI circuits are developed and as the cost of the remainder of the exchange decreases. It is also desirable as it creates a network poised to adopt ISDN (Integrated Services Digital Network) techniques when the analog-to-digital interface will be transferred to its ultimate position at the subscriber's premises or the telephone instrument.

The multiplexing technique immediately following coding can already be used to introduce a measure of concentration by multiplexing more than 32 lines onto each 32-channel system. Multiplexes of 64 and 128 lines per 32 channels are commonly used. The limit to the concentration applied at this stage is one of reliability as the size of the group of subscribers dependent upon a single non-replicated equipment of considerable complexity must be limited.

The first commercial digital local exchange system (and today the system with the most lines in service), the E10 System of CIT Alcatel, France, employs analog concentration stages prior to analog-to-digital conversion. While developing a digital concentrator, CIT continue to give cogent reasons for retaining

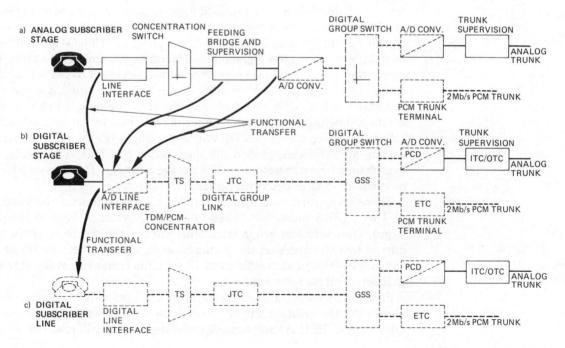

Fig. 5.20 Transition stages towards ISDN: AXE system

analog concentrators for all but the (relatively few) subscribers requiring a digital interface to the ISDN. L M Ericsson of Sweden developed their AXE local system with analog concentration but now provide digital concentration.

Figure 5.20 illustrates the transition through successive generations of the AXE system from analog concentration to the ISDN.

The AXE 10 Subscriber's Switching Subsystem

The AXE system will serve well as an introduction to the functions of the subscriber's concentration stage of a PCM local exchange. *Figure 5.20* illustrated that, as analog-to-digital conversion is moved to the periphery of the exchange, many other functions are moved to the line interface also. The resulting line interface is shown in functional terms in *fig. 5.21*. These functions will be discussed more fully in Chapter 8; it suffices now to indicate that the elements of the mnemonic BORSCHT are evident in *fig. 5.21*:

- B   Battery feed
- O   Overvoltage protection
- R   Ringing
- S   Supervision (detection of seizure and release) or Supervision and Signalling
- C   Coding
- H   Hybrid or 2-wire to 4-wire conversion
- T   Test

The principles of subscriber switching in the AXE system are illustrated in *fig. 5.22*. Subscriber modules of 2048 subscribers are divided into sub-modules of 128 subscribers, each with its own time switch, key sender receivers, test circuits and an optional 2 Mb/s link to the group switching stage. All 16 sub-modules have access to a common time switch bus having 512 channels. Thus, connection to the group stage is effected via a channel of the sub-module's 2 Mb/s link or, if all these are busy or the link is not fitted, via one of the 512 channels linking to other sub-modules over the time switch bus. There is thus a time division bus interconnecting the lines within each sub-module called the device speech bus DEVSB and a single 512-channel bus interconnecting all 16 modules called the time switch bus TSB. The speech store in each time switch therefore has 768 storage locations. This is more adventurous than the devices described earlier in this chapter.

There is capacity, therefore, to provide a normal minimum concentration of 4 to 1 (2048 subscribers' lines to sixteen 32-channel links to the group stage). Only sufficient group stage links will be provided to carry the traffic offered by subscribers on the particular exchange and, because of the time switch bus (TSB) link, traffic from all subscribers has full availability to all channels of all the links fitted.

Figure 5.23 shows a block diagram of the time switch in concept only, illustrating the relationship of channel capacities of the various inlets to DEVSB and TSB. At any time-slot, the time switch will read out a sample

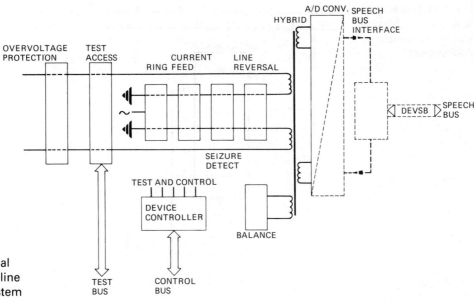

Fig. 5.21 Functional diagram of digital line interface: AXE system

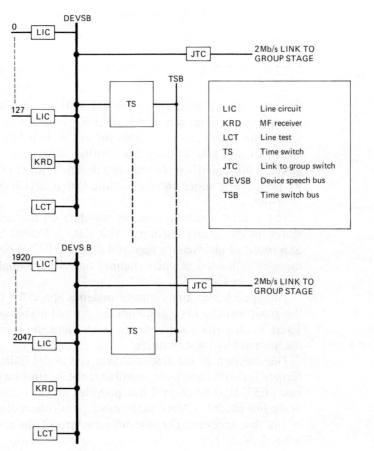

Fig. 5.22 Principle of subscriber stage switching: AXE system

LIC	Line circuit
KRD	MF receiver
LCT	Line test
TS	Time switch
JTC	Link to group switch
DEVSB	Device speech bus
TSB	Time switch bus

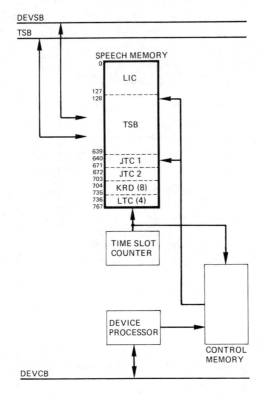

Fig. 5.23 Functional diagram of subscriber stage time switch: AXE system

from the speech store on to the DEVSB bus destined for a subscriber or a channel to the group switch or a VF receiver or to line test *and* at the same time it will read a sample on to the TSB bus destined for the group switch or for a subscriber via another subscriber stage time switch. The AXE 10 subscriber's stage time switch is therefore performing space switching by performing more than one time-slot interchange at the same channel time-slot.

The control memory contains the data on the destination of the sample stored in the speech memory. This data is loaded by the device processor as a result of instructions received on the DEVSB bus. Note that the speech memory is loaded at each channel time by a sample from DEVSB *and* a sample from TSB.

The speech memory capacity includes space for two 32-channel links to the group switch, JTC, although the normal maximum is one JTC per LSM. There is also spare capacity for more than the normal 8 VF receivers and the normal 4 line test facilities.

The diagram of the line interface (*fig. 5.21*) indicates that there are two further time division buses involved; the control bus DEVCB, and the test bus TEST B. The control bus provides access from the regional processor of the line module EMRP to the device controllers of the constituent elements of the line module. The control structure of the line module is illustrated in *fig. 5.24*.

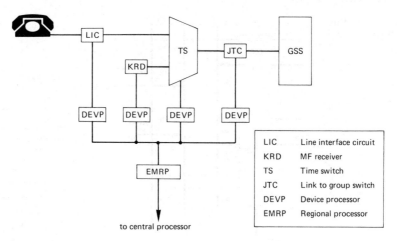

Fig. 5.24 Control structure of line module LSM: AXE system

LIC	Line interface circuit
KRD	MF receiver
TS	Time switch
JTC	Link to group switch
DEVP	Device processor
EMRP	Regional processor

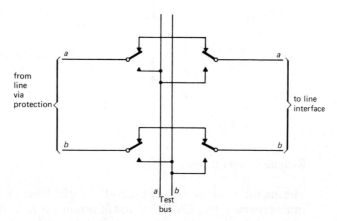

Fig. 5.25 LIC test access relays

The provision of solid state electronic switching right out to the subscriber's line interface, particularly in TDM form, makes physical access to the subscriber's line for testing purposes impossible. Line and trunk testing is a subject that will be returned to in Chapter 8. At this stage it is necessary to point to the need to provide other means, in TDM exchanges, to gain access to the lines. The access method employed has become almost classical in its universal adoption. *Figure 5.25* illustrates the contents of the LIC block labelled Test Access in *fig. 5.21*. Test access relays are provided which normally (when released) provide a through connection from the line to the line interface. Operation of one test access relay connects the line to a test bus for outward testing, whereas operation of the other relay provides access from the test bus to the line interface for inwards testing of the interface and other exchange functions. In some systems (notably ITT System 12) the relays are also used to switch in a spare line circuit in place of a faulty line interface.

The interface arrangements with these three bus systems, DEVSB, DEVCB, and TEST B, are shown in *fig. 5.26*. This introduces a point, expanded in the next chapter, that time division relates, not only to the circuit switching function, but also to the control functions of the telephone exchange.

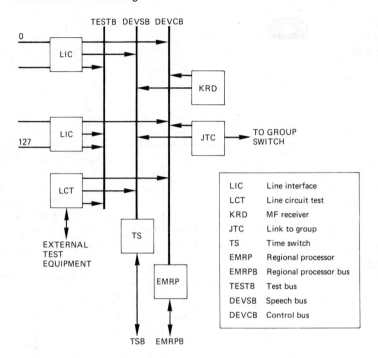

Fig. 5.26 Line module highway (bus) interfaces

Remote Concentrators

The migration of the bulk of the cost of a telephone exchange to the periphery, caused, in part, by PCM TDM and illustrated in *fig. 5.20*, has been encouraged to proceed still further. One aspect of this is the advent of ISDN where the migration reaches the subscriber's instrument which becomes a communications terminal and work-station as a result. Another, less drastic aspect is the attractiveness of locating the subscriber's concentration stage remote from the telephone exchange. There was always a requirement for remote concentrators and suitable space division equipments were designed, mainly for rural application. The costs involved never justified such concentrators in more urban locations since a reduction in the local cable of between 80% and 90% (one link to the exchange for every 10–20 subscribers instead of one per subscriber) could not be made to pay for the cost of the remote switch plus its secured power supply and ancillaries. Added to this was the extra maintenance effort involved in routine maintenance of remote electro-mechanical devices.

With PCM TDM, the reduction in links to the main exchange is much greater (two *channels* for every 10 to 20 subscribers), the power requirement is lower (although seldom low enough for power to be supplied down the line from the parent exchange), and the necessarily complicated signalling can be accommodated on common channel signalling links. Almost all practical PCM local switching systems therefore include remote multiplexor, remote concentrator and, possibly, remote simplified exchange options. Because of TDM, these options can also accommodate diversity of routings to the parent exchange, or even exchanges, thus overcoming the other disadvantage of concentrators that service is lost to large numbers of subscribers should the exchange link be broken. At the least, concentrators with just one link to the parent continue to provide local service when that link is broken. Because the concentrator is cut off from the processing power of the parent exchange, this residual local service is probably not charged. This is small comfort, however, to the subscriber, accustomed to access to the worldwide telephone network but now reduced to access to only a few hundreds of nearby subscribers.

The AXE 10 concentrator is similar to the subscribers switching stage just described except that the links to the group stage now become 30-channel CEPT systems, and the control links from the regional processor to the central control are formed by using common channel signalling via time-slot 16 of two of the 30 channel links to the parent exchange. This concentrator arrangement is shown as part of the diagram of a full AXE 10 local exchange in *fig. 5.27*. The diagram illustrates that the only difference in the remotely located subscriber stage is the use of common channel signalling to communicate between the remote regional processors and the exchange central processor.

Channel Modularity and Subscriber Signalling

In this brief survey of a practical subscriber's switching system several concepts have been introduced without detailed discussion. One, which is evident from the diagrams, is that a modularity of 30 channels has been increased to 32. Similarly, the diagrams reveal units devoted to signalling: KRD (MF receivers) and ST-C and ST-R (common channel signalling equipment).

Channel Modularity In describing the 30-channel system, channel 0 was identified as being reserved for synchronisation, frame alignment and other link related functions, and channel 16 for signalling. However, these functions, possibly required outside the exchange, are certainly not necessary within the exchange. Most practical switching systems, therefore, utilise all 32 channels for traffic-carrying connections. Channel 16 is identified in the CEPT system as a signalling channel and a quartet system is defined for channel associated signalling. It is often appropriate, therefore, to utilise channel 16 as a first choice for common channel signalling in which case the quartet protocol is abandoned and the channel becomes a normal communications channel of 64 kbit/sec capacity. This is the arrangement chosen for the AXE 10 system. Throughout the book there is a distinction drawn between channel 16 signalling (channel associated in quartets) and signalling over channel 16, normally probably common channel but on occasion the channel may be used for ordinary traffic-carrying circuits. As an example, were the remote concentrator of *fig. 5.27* to require 3 or more "30"-channel links to the parent exchange, then the remaining channel 16s could be devoted to ordinary traffic.

96 Introduction to Digital Communications Switching

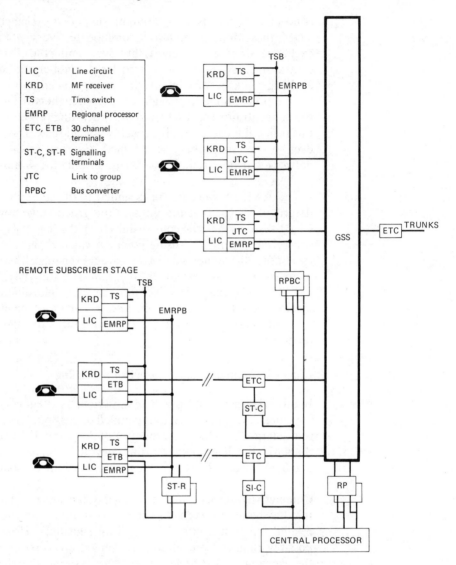

Fig. 5.27 AXE 10 local exchange with exchange located and remote digital subscriber stages

LIC	Line circuit
KRD	MF receiver
TS	Time switch
EMRP	Regional processor
ETC, ETB	30 channel terminals
ST-C, ST-R	Signalling terminals
JTC	Link to group
RPBC	Bus converter

Subscriber Signalling Signalling from the subscriber to the exchange is normally loop seizure and release, with either loop-disconnect dial impulses or multi-frequency key-sender signalling for routing information. In either case the digital exchange must provide means for detecting the routing request information and passing it on to control. If loop-disconnect is used then this can be detected, in the PCM line circuit scanning process, by (in the case of AXE 10) the LIC device controller and passed on to the regional processor. MF signalling will pass through the line circuit and be PCM-encoded. There must, therefore, be a MF receiver associated with the line circuit when it is seized and a TDM connection completed to this receiver which detects the digitised MF information, translates it and passes the resulting numerical information to the regional processor. Chapter 7 deals with this topic in more detail.

Chapter Summary

After four chapters of introductory material, we arrived at the central subject of digital communications switching, but beginning with a *theoretical* treatment of three-stage space-division switching networks. This led on to the main discussion of the requirements of time division switching establishing that the need to store samples in time switching indicates the desirability of using digitally encoded samples. The time switch function and the space switch function were then discussed in some detail, with the time switch presented as a mechanism for time-slot interchanging.

The time switch and space switches were then brought together into switching arrays combining both functions, thus making clear the relevance of the earlier treatment of three-stage space arrays in terms of gaining some measure of the efficiency of the PCM TDM approach.

The treatment adopted up to this point in the chapter owes a considerable debt to John Bellamy [see Reference 5.1].

TDM switching is a significant source of delay and this aspect was discussed prior to completing the treatment by reference to a modern practical example. The example system (AXE 10) was used to demonstrate the approach to switching in the subscriber's concentration stages of the switching process. The absence of continuous metallic connection to the subscriber's line introduces the need for a new means of line test access which was also discussed. Lastly, the provision of remote subscriber's switching, a feature made conveniently possible by PCM TDM, was outlined.

Reference

[5.1] *Digital Telephony*, John Bellamy (John Wiley and Sons, 1982). This excellent work has been used extensively in the early part of Chapter 5 and has also been of great assistance in the material of Chapter 4.

Exercises 5

5.1 Give reasons why, historically, the first digital switching systems placed in service were trunk switching systems.

5.2 Comment on the reasons for the difference in figures for the cost distribution of analog and digital exchanges shown in *fig. 5.2*.

5.3 Draw a three-stage array which connects 400 inlets and outlets using 20 inlet switches in the first and third stages. The array is to be non-blocking.

5.4 Suggest a more efficient array to perform the functions required in Exercise 5.3.

5.5 Calculate the missing figures in Table 5.2.

5.6 A four-wire connection is made between Channel 3 System M, Channel 3 System X (return), and Channel 11 of Systems N and Y (see *fig. 5.8*). If Systems N and Y form the far end of the connection, what is the minimum round trip delay of the connection for a signal initiated via Channel 3 System M to cause a response at Channel 3 System X?

- **5.7** Give reasons why time division switching should be PCM TDM switching.
- **5.8** Using a diagram such as *fig. 5.11a* demonstrate how incoming channels 2, 7, 15, 21 are switched to outgoing channels 17, 1, 7, 8 respectively.
- **5.9** Draw a diagram illustrating the processes occurring in a TSTST switching array. Comment on the practical utility of such an array.
- **5.10** Discuss the relative merits of TST and STS arrays.
- **5.11** Discuss the relative merits of stage-by-stage switching and link frame switching. Speculate on the reasons why methods of end-to-end route choice through exchange switches have appeared only relatively recently.
- **5.12** Why would particular subscribers experience a much worse service than that provided by the exchange overall? What steps can be taken to improve the situation?
- **5.13** Write twenty-word sentences explaining each of the letters of the mnemonic BORSCHT.
- **5.14** An AXE 10 system has 11 JTC links to the group stage. What is the concentration ratio for a fully equipped LSM co-located with the exchange? What would the concentration ratio be if the same LSM were remotely located?
- **5.15** Draw a diagram indicating how the test access relays of *fig. 5.25* could be used to switch in a spare line circuit instead of a faulty line circuit. Comment on possible snags in the approach.
- **5.16** List the advantages and disadvantages of remote concentrators.

6 Digital Exchanges and Control

Introduction

The subject matter of this chapter is shown by the shaded portion of *fig. 6.1*. The control function was the first part of the exchange to employ time division switching. Even the early marker-controlled crossbar systems demonstrate a measure of time-divided common control as the same control machine was used to perform a control function for a number of different call attempts in a time-divided ordered process. To a very high degree in fact the content of this chapter is very little influenced by the advent of digital communications switching. The control task, though changing in detail to deal with TDM switching, is, in essence, the same as the control task in space division switching.

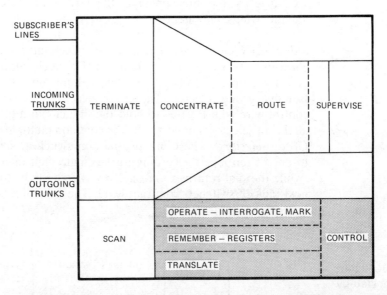

Fig. 6.1

Marker-controlled crossbar systems were, then, the first systems to use a measure of common control. The degree of centralisation of the control was limited by the speed of the markers using, as they did, electromechanical relays. It was perhaps natural to assume that the increased speed of electronic devices would allow further control centralisation and that this would be desirable. Recent trends have, to an extent, reversed this process.

There is a tendency to describe the control of communications switching as a task suitable for a computer and indeed much of mainframe, mini- and micro-computer technology and software technique are applicable to exchanges. There are two crucial differences in the application however. The telephone exchange is expected to operate 24 hours a day for 20 years without complete failure and the telephone exchange operates in real time. A further

100 Introduction to Digital Communications Switching

difference of less importance is that control of a communications switch does not require much arithmetical manipulation.

In this chapter the subject is approached by means of a number of routes, outlining the problems and functions involved on the way. One such route is to follow the historical tendency towards greater centralisation of control. Another is to outline the various methods of providing security for centralised (or any form of) control. Emerging from these themes will be a third, leading from the modern tendency to distribute control either by functional division or by a more radical distribution of all control concepts, to a consideration of the software functions of control.

The Trend towards Centralisation

Step-by-Step

The earliest automatic telephone system, the Strowger step-by-step system, is still the finest example of **distributed control** (fig. 6.2). In this system each switch is equipped with its own control equipment. *Figure 6.2* illustrates an exchange as it would appear in the UK non-director system consisting of the following components:

Subscriber's line circuit and uniselector Each subscriber "owns" a line circuit consisting, typically, of two relays and a single motion switch. The relays detect a loop and cause the switch to hunt over its 25 or 50 outlets for a free first group selector. The group selector returns a condition on a third control wire (the P-wire) to hold the line circuit and to busy that outlet on all the uniselectors connected by the common multiple to that group selector.

Group selectors These are two-motion switches, stepping their wipers up to one of ten levels according to the single digit dialled and then searching (while the dial returns to normal) for a free outlet (one of 10 or 20) to the next rank of switches from that level. The control relays of a group selector

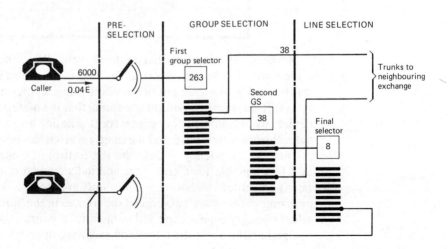

Fig. 6.2 "10 000-line" non-director step-by-step exchange

are typically four in number: A, impulsing; B and CD, recognition of dial interdigit pause and hold during dialling; and H, switching. There will be as many ranks of group selectors as there are digits to dial less one; the last two digits are used by the final selector.

Final selectors These use exactly the same mechanism as the group selectors but step up for the penultimate digit and into the bank for the final digit, thus identifying the particular wanted subscriber's line. The final selector also, usually, contains the feeding bridge for both subscribers (providing DC feed to both subscribers but only an AC path between subscribers). The final selector, therefore, detects release and removes the P-wire conditions to release the complete switch train.

A group selector is shown in *fig. 6.3*. There are added complications associated with trunks to other exchanges but this simple description will suffice for our purposes. (Reference 6.1 is recommended for the interested reader.) *Figure 6.2* actually shows less detail than *fig. 1.11* from which it is derived, so that a glance at *fig. 1.11* may help.

To provide an approximate idea of scale, the exchange of *fig. 6.2* has been dimensioned very roughly assuming that subscribers originate $0.04\,E$ each and the probability of loss at each stage is 0.01. This ignores more complex problems of limited availability switches. For a 6000-line exchange then, the approximate numbers of switches required are as follows:

Subscriber's line circuit and uniselector	6000
First group selectors	263
Second group selectors 7×38 (level 1 and level 9 not used)	266
Final selectors $7 \times 9 \times 8$	504

The heading of *fig. 6.2* uses inverted commas for good reason. Theoretically a 10 000-line exchange, the system shown is actually capable of far less. Levels 7 and 74 are trunked to neighbouring exchanges reducing the capacity of this exchange by 1000 and 100 lines respectively. BT (and the Post Office before them) never used first selector level 1 for subscriber connections because a 1 can be so easily imitated by flashing the switchhook on the telephone (while dusting, for example). Thus the maximum subscriber capacity of the exchange shown is 7899 lines. This illustrates a major problem of step-by-step systems: the exchange numbering system is determined by the switching system. Let us consider another facet of this problem.

Figure 6.4a illustrates several exchanges in the same area. Some quite clever design work has to be devoted to devising a suitable numbering scheme. Some exchanges are potentially bigger than they need be because of the numbering; and all this design has to be re-arranged because of the need for a new exchange in the area. Subscribers are confused by having their numbers changed, apparently quite arbitrarily. As shown, subscribers on exchange D cannot dial through A to C unless the numbering is re-arranged as shown in *fig. 6.4b*. In this case a different number has to be used when dialling C from A to that required when calling from B, and the numbering string has been

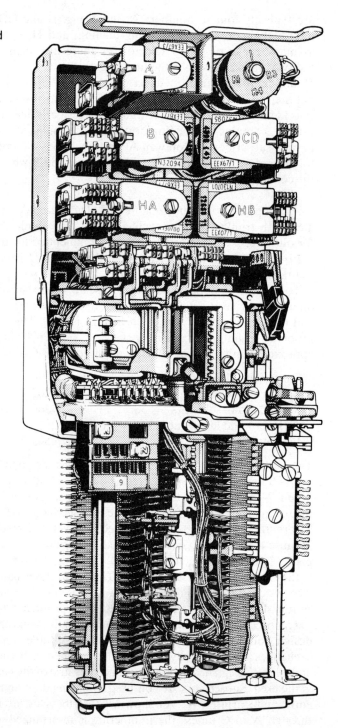

Fig. 6.3 A two-motion type selector equipped as a group selector

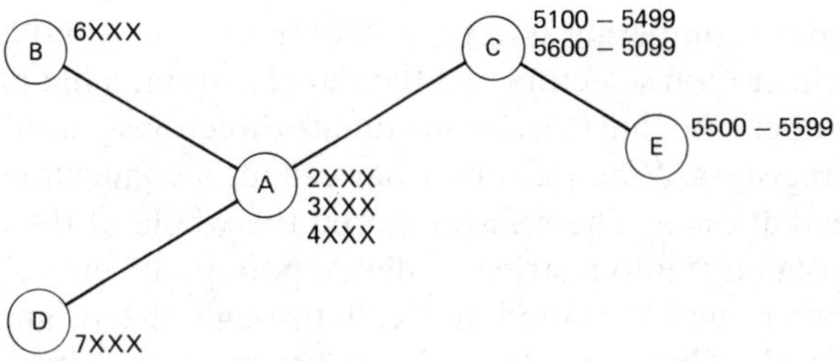

Fig. 6.4a Strowger area numbering: inadequate numbering (D cannot dial C through A for example)

Fig. 6.4b Strowger area numbering: adequate mixed numbering

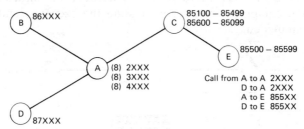

Fig. 6.4c Exchange A trunking: providing the requirements of fig. 6.4b

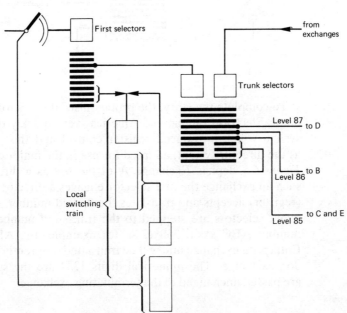

increased in most cases from 4 to 5. This increase is to accommodate the routing requirements. The dilemma is illustrated forcefully by *fig. 6.4c* which shows the trunking of Exchange A necessary to provide the requirements of the numbering shown in *fig. 6.4b*.

This was the difficulty that faced the British Post Office (BPO) in the 1920s when they wished to choose an automatic system which would be standard for the whole country including the major conurbations where (in the London area particularly) up to 400 exchanges were required to be interconnected in a common numbering scheme. Clearly this is impossible using the Strowger system as described above.

Other competing systems at that time—the Rotary system of common drive single motion selectors and the Panel system, a flat panel of contacts accessed by wipers moving in two co-ordinate directions—used systems of control based on registers. When a subscriber called, a connection to a register was established at once. The register stored the whole of the dialled number and then translated it into a string of digits, potentially entirely different to the dialled number, which routed the call through the originating exchange *and the network*. Thus, the dependency between the directory number dialled and the equipment number expressing the exchange and network address was eliminated.

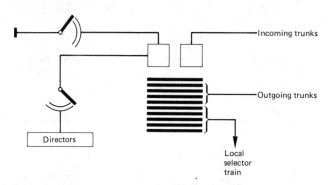

Fig. 6.5 Exchange A as a director exchange

To complete the story, the protagonists of the Strowger system, The Automatic Electric Company of America, very quickly devised a version of the system which introduced register control and this was the system, known as the director system, adopted for use in the major conurbations in the UK. *Figure 6.5* depicts Exchange A of *fig. 6.4* as a director-type exchange. In such an exchange the first selector engages a director on seizure, the director (register) accepts and translates the dialled number, and the first and subsequent selectors are stepped to the translated number. Thus, for a London number ABC xxxx, 368 1234, for example, the ABC digits, 368, identify Enterprise exchange and will be translated differently in each originating London exchange. The numerical digits 1234 are the subscriber's number and are passed unchanged to the terminating exchange.

The Lessons of Step-by-Step

The distributed nature of the control in the Strowger system makes it pre-eminent in reliability. A multiplicity of available paths through the switch, each path controlled separately, means that failure of a control element disables that path only, never the exchange. The most serious risk of failure in the system shown in *fig. 6.2* is the failure of a final selector. In such an event the 100 subscribers affected may still be accessed via one of seven other final selectors. This excellent security is paid for however. The control elements are used only as the call is being set up and disconnected, a total duration of perhaps half a second in a call averaging 100 seconds. Clearly, this is very wasteful in the use of expensive control devices.

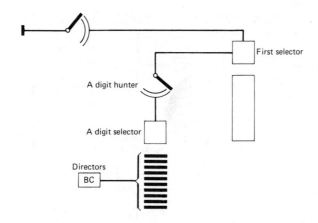

Fig. 6.6 UK director system

The modification of the Strowger system to provide register control, the director system, had already moved in the direction of centralisation, and the directors became the area of least security. To overcome this and to provide the level of service required with slow-speed electromechanical devices, the director was sub-divided into directors specialised for each hundred combinations of B and C digits. Further all directors were replicated on a traffic basis. *Figure 6.6* shows the UK director system in more detail.

This description of the step-by-step system introduced the P-wire, a third wire connection from end to end of the exchange set up in parallel with the conversation connection and used for control and signalling between the various exchange modules. The P-wire or its equivalent is a necessary part of any exchange connection using any system and will reappear in various forms in all the systems to be discussed.

Towards Centralised-control Marker Systems

The traffic efficiency of switches with an availability of 10 or 20 is not particularly good. The traffic efficiency of a system which chooses free outlets with no knowledge of subsequent busy conditions is still worse. The availability consideration indicates the use of switches of higher availability (more outlets) and this alone leads us towards link systems and overall choice of path across parts of the network (the frame) or across the complete network. To achieve this involves departing from the undoubted security of Strowger to employ a measure of **centralised control**.

Early systems employing centralised control and register control (rotary, panel and crossbar) were limited in the degree of control possible by the speed of the devices. We therefore saw a generation of link frame systems employing marker control of the frames and register control of the routing. *Figure 6.7* illustrates an example of such a system in its fully developed and very sophisticated form.

The move away from the distributed control of Strowger to the director system or to link frame systems introduced new problems. Chief among these, and obvious in *fig. 6.7*, is the need for low capacity, high availability, high

Fig. 6.7 Register-controlled link system: ITT Pentaconta

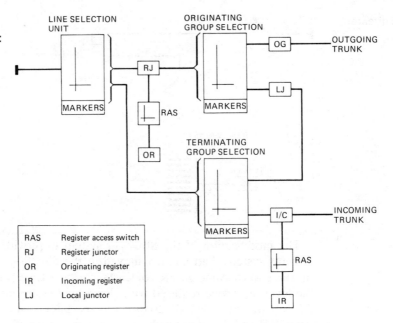

security and, therefore, expensive switches to access the registers. The complexity of the network design problem, solved by register control, resulted in shifting the problems into the register and associated equipment. To modify the ITT Pentaconta system for UK use took almost five years of intensive design effort in the early 1970s to modify the register.

Control Security

We now turn to the second theme of approach to the exchange control problem, that of security. We have seen the disadvantages of a fully distributed control structure in the relatively simple environment of the step-by-step system. We will now assume that an exchange switching system is required which has fully centralised control. In such a system (*fig. 6.8*) there will still be registers but their function will be confined to an unthinking receipt and storage of digits with reference to control for instructions on what to do as a result of receiving a particular digit stream. *Figure 6.8* represents no particular actual system although many actual systems are very similar.

In *fig. 6.8* can be seen all the functions required of the central control:

Scan control Causes all subscribers' lines and incoming trunks to be scanned regularly looking for new call requests.
Register control Receives dialled data from registers; determines exchange and network routing; provides register with network routing data for onward sending.
Switch control Sets up required paths through network. Note that this includes an initial path to the register on first detecting a new call.
Housekeeping Charge determination, detection of faults, keeping traffic records, etc.

Fig. 6.8 Concept of common control switching system

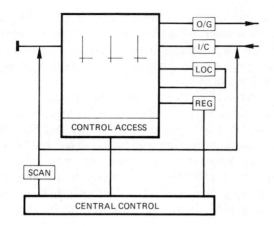

Clearly the operation of the exchange depends entirely on the central control and steps must be taken to ensure that it does not fail or, if it fails, that a reserve control can take over the duty. There are various ways of solving this problem, at least in part, and these are illustrated by the series of diagrams of *fig. 6.9*.

If a single control cannot be trusted to operate without failure for 20 years then a fairly obvious solution is to provide two such controls. The problem then arises first of detecting failure and then of substituting a new control with sufficient data on the state of affairs in the exchange to assume duty.

Duplicate Synchronous Control

Figure 6.9a illustrates duplicate, synchronous control: two identical machines working in synchronism so that they each perform the same task at the same time down to the minutest detail. It is therefore possible to compare their output step by minutest step and cause an alarm should the outputs disagree. Fault detection is therefore very easy. But fault correction is difficult because no evidence is available as to which output was correct and therefore which control should be disabled.

One method of solving the fault correction problem is to add a third control in micro-synchronism with the other two. In this case majority decision voting will easily identify the faulty unit. Alternatively, micro-synchronism of control

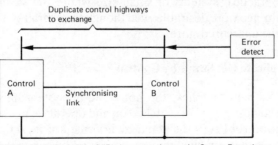

Fig. 6.9a Duplicate synchronous control

CONTROL CAPACITY = 2 (Exchange requirement) + Sync + Error detect modules (smaller sub-sections of the control machine) can be used. The modular approach allied to the use of error-detecting codes on all outputs from control and all interfaces between control modules has been employed in practical systems. A sophisticated version of this approach is used by the Ericsson AXE 10 system.

It is plain therefore that the concept illustrated by *fig. 6.9a* requires considerable elaboration in order to realise a practical system. This is true of all the alternatives that we shall consider. It is clear also that the method leaves a residual amount of equipment that is not secured by duplication. This residue is, however, very simple, consisting of error detection by comparison and synchronisation, probably by locking duplicate pulse generators together. The additional equipment required to detect which of the two was the faulty control would be included as part of each control and therefore secured by duplication.

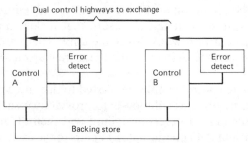

Fig. 6.9*b* Duplicate cold stand-by control

Duplicate Cold Stand-by Control

Figure 6.9b illustrates duplicate control where the stand-by control is normally switched off. The storage associated with any control scheme is, in this case, kept separate and is available to both controls. Again error detection is duplicated and is performed on redundantly coded messages from control. In this case, on failure of the working control, the stand-by is switched on, reads its programs and data from the backing store, and then proceeds to run the exchange. There is, therefore, a hiatus period during which no control of the exchange is possible and therefore calls in progress will be lost.

Because of the danger of losing calls this approach has only been mooted for practical systems of small size where the savings in power and savings in storage were desirable even though the overall system security is low because of the unsecured store.

Duplicate Hot Stand-by Control

As its name implies, the hot stand-by system illustrated in *fig. 6.9c* keeps the stand-by control switched on and operating in a stand-by mode. The working control keeps the stand-by informed of its actions via the interprocessor link so that, on change-over, the stand-by can take up duty immediately and

Fig. 6.9c Duplicate hot stand-by control

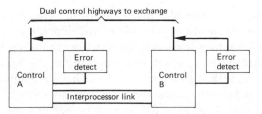

CONTROL CAPACITY = 2 (Exchange requirement) + Work to keep stand-by informed

calls in progress are not lost. The interprocessor link is a more complex, common, unsecured element requiring very special and expensive provisions to limit danger of complete failure due to the link. And the controls now have additional complexity to handle the interprocessor communication.

Dual Load Sharing Control

All the options so far discussed require the provision of a second control in addition to that which is operating the exchange. The investment in security is at least equal to the investment required to operate the exchange. The hot stand-by solution provides a clue to a new approach. With the stand-by control switched on and working, why should it not be used to control the exchange in parallel with the working unit? Were this the case, the stand-by could at least help with over-load situations and it might be possible to reduce the processing power of both machines so that exchange operation, though maintained upon control failure, would be maintained at a less than optimum level.

Fig. 6.9d Dual load sharing control

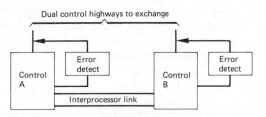

CONTROL CAPACITY = Exchange requirement + A + B

where A = work required to inform other processor
B = allowance for additional work when only one processor is operative.

This is the idea underlying all the arrangements yet to be discussed and is particularly the concept of dual load sharing illustrated in *fig. 6.9d*. This figure is in fact identical to the preceding diagram but the controls themselves would be dimensioned, in storage capacity and throughput, so that one machine catered for the exchange capacity at a level less than optimum but both machines would give greater than optimum performance. Exchange operation is entrusted to both machines and they each keep the other informed of their activities via the interprocessor link.

This is the architecture employed by the ITT Metaconta range of stored program controlled exchange systems. It was continued by the Thomson CSF organisation in their MT range of digital switches. (Thomson inherited the concept when they bought Le Material Telephonique from ITT.) The architecture has therefore had some 20 years of useful life in practical SPC analog and digital systems, which is a long time in this modern age of SPC where fashion rather than technical excellence has sometimes been the rule.

Dual load sharing control requires considerable additional processing power in each control devoted to the task of keeping the other control up-dated on its activities. Provided this up-dating process is satisfactory it is possible for either control to deal with each request for processing rather than the complete call process. On failure therefore all calls, including calls in the process of being completed, can be processed.

Multi-processing Control

All the architectures of control so far discussed have been based upon a single large machine, very similar to a mainframe computer, duplicated in some way for security. The technical trend encouraging this approach was undoubtedly the desire to use stored program control (SPC), where the logic of exchange control is provided by software programs stored in the control memory. Previously the control logic was stored in the physical connections of the logic circuits. The real-time traffic-dependent nature of the control task does indicate that an alternative, possibly better, approach to the problem would be smaller, simpler, distributed controls utilised and provided on a traffic basis. Some few systems, notably the UK TXE 4 system, swam against the SPC tide towards big control machines and used this approach very successfully. Eventually technology turned the whole industry towards smaller distributed control architecture by providing the integrated components to design mini- and micro-computers and, today, single-chip processors.

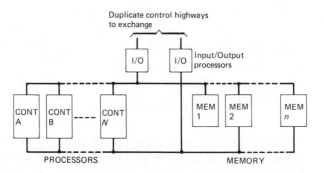

Fig. 6.9e Multi-processing control

One tentative step towards distributed microprocessor control named *multi-processing* is illustrated in *fig. 6.9e*. The control machine has been sub-divided into component parts: control processors of identical design provided on a traffic basis to handle the busy hour call attempt (BHCA) load and intercommunicating between themselves; the memory (also sub-divided); and the input/output processors via high-speed high-security bus highways. The input/output processors provide the processing power and storage capability to "change gear" to the speeds and protocols necessary to communicate with the exchange devices.

In a system using multi-processing control, indeed in any modern system claiming to use distributed processing, the rest of the exchange will have a considerable additional processing power with small processors or controllers in each unit receiving, interpreting and implementing control commands and collecting and sending information to control. With this arrangement, control capacity can be closely geared to exchange traffic but there is a limit to the maximum multi-processor configuration, set by the speed of the high-speed bus highways. Modern systems often require BHCA capacities well in excess of this limit and it is necessary to control the larger exchanges using several such multi-processor clusters. Each new cluster carries significant overhead of highways, input/output and inter-cluster communications. Multi-processing is the approach used in the UK System X digital switching system.

Distributed Control

As has been mentioned, fashion has not been absent from the tendencies of exchange control architecture. "Fashion" at the present time spells distributed control by means of multiple microprocessors. It is therefore an additional confusion to the student that some of the control architectures already described are categorised by their manufacturers as "distributed control". The basis of the claim rests on the proliferation of processors controlling the various functional modules of the exchange.

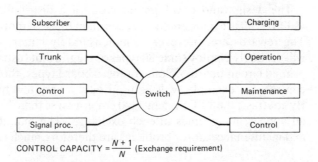

Fig. 6.9f Distributed control

CONTROL CAPACITY = $\frac{N+1}{N}$ (Exchange requirement)

The concept illustrated inadequately by the diagram of *fig. 6.9f* and described as distributed control is something rather more specific than centralised control of distributed functions themselves controlled by microprocessors. The concept is best illustrated by using the same argument that was used in its development.

Digital switching, with digital encoding at or close to the subscriber, re-distributes the cost of the exchange, as we have seen, and also re-distributes the signalling capability. There is available high capacity voice, data and signalling means direct to the subscriber. To use this high capacity each exchange terminal (subscriber) requires a processor. While recognising that this happy state of having a processor per termination had not yet arrived, the designers of ITT System 1240 argued that this was the trend. They therefore postulated an exchange controlled by multiple processors distributed to the periphery of the switch and intercommunicating by means of the switch.

Each group of lines (64 or 128) and each 30-channel trunk group, each register (now known as MF sender/receivers with the advent of push-button multi-frequency dialling), and each exchange function (charging, termination status, routing, maintenance, etc.) is controlled by one or several of a large number of identical microprocessors. A single call connection request may involve the cooperation of three to six such processors each performing different functions of the call connect process. In fact, a voice call connection is only one of many connections made by the switch as it is continually interconnecting microprocessors so that they can communicate in carrying out their different duties.

Such a control architecture is only made possible by utilising a very special design of self-seeking digital switch which enables any processor to launch a message into the switch containing the address of the wanted correspondent processor. The switch itself acts on the address to connect the message. Rather like Strowger? Yes.

Some of the snags and pitfalls of distributed control will not become evident until we have discussed switching software architectures. At this point we will note only the delay involved in controlling the exchange by means of a multitude of intercommunicating processors. The time overhead of the many messages involved can be substantial and necessitates very careful design of the software modules.

The design and development costs of a distributed control system such as ITT 1240 are undoubtedly greater than those of more conventional systems. The rewards are, however, commensurately larger. A conventional system will probably employ some 80 types of plug-in unit (usually single or multi-layer printed circuit boards). Some of these board types, those used in the duplicate controls for example, will be used only a very few times in each exchange. By contrast, the ITT System 1240 employs less than 30 board types and nearly all appear many times in each exchange. The manufacturing, maintenance and maintenance stores problems are therefore much reduced.

Control Software   The discussion so far has often approached, and then veered away from, the subject of control software. It is now high time to deal with the subject at least in outline. It is not impossible to argue that "software" was expressed by the wiring of even the earliest forms of exchange control. The difference between "wired software" and electrically stored software impinges for the most part upon the system design process. A wired software system would not work until all its constituent parts had been manufactured. The logic of the system was designed at the same time as the system operating "hardware" (switch, store, detector, frequency sender, coder, decoder, etc.). With stored program control the two design processes can be separated and software design can (in theory) continue while the hardware of the system is being manufactured. SPC systems also (it was claimed) can accommodate changes and upgrades more readily.

In real life developments, particularly developments lacking impeccable management, the undoubted separation of the logic into software and the execution into hardware has resulted in the software design proceeding while the completed hardware stood idly waiting on the exchange floor. This has resulted in instances of software with so many "patches" (on-the-spot changes to make the software work) poorly recorded, that the final state of the software is not known. In this situation upgrades become impossible.

To introduce the concepts of software we will analyse the processes of completing a call through a telephone exchange using some of the concepts first introduced in the latter half of Chapter 1.

Call Progress

Figure 6.10 depicts the processes involved in the lifetime of a single call from a subscriber on an imaginary exchange system to a subscriber on another exchange somewhere else in the network. This is a complex diagram which probably deserves a pause for close study. The diagram divides the exchange world of events into three sections: *telephone events* or actions by the network upon the exchange; *exchange actions* or actions by the exchange executive equipment; and *actions by control* or actions requiring logical process. We have further indicated a division in time as the call progresses through pre-selection, selection, call completion and release, concepts introduced in Chapter 1.

Figure 6.10, though grossly oversimplified, already indicates major functional differences between, say, pre-selection and selection. In writing software for a new system, therefore, it would probably be wise to define PRE-SELECTION, SELECTION, CALL COMPLETION and RELEASE as separate modules.*

Not quite so evident in *fig. 6.10* but indicated nevertheless is the concept of data. The control instruction, "connect ring tone to sub", will contain a network address for ring tone and a network address for the subscriber. These addresses are data, established at exchange opening and seldom changed. The subscriber network address is associated with a subscriber directory number (DN) and a list of the special features of the subscriber. This data (DN and subscriber features) is established at exchange opening and subject to relatively frequent change. The subscriber dials a number for the party required; this is data input at the start of a call and constant only for the duration of the call. In writing software for control of communications switching systems (or for any other purpose), it is essential to maintain a rigid separation between the program (the logic) and the data. This is not always as simple a matter as it might seem.

---

* Software designers seldom seem to leave room in their coding fields for decent English so that these modules might rejoice in names such as: PRE, SEL, CACOM and REL.

Fig. 6.10a Call progress

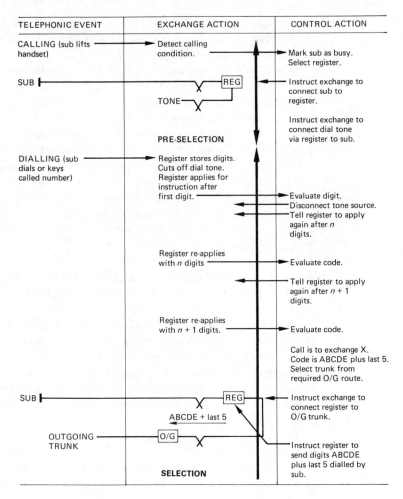

Software Modularity

In discussing *fig. 6.10* we have identified different areas in the call connection process, recognised a difference in kind between software logic and software data, and seen a difference in the nature of the data. We can summarise these findings as follows:

SOFTWARE MODULES
Call connection   Pre-selection
                            Selection
                            Call completion
                            Release

Fig. 6.10b Call progress

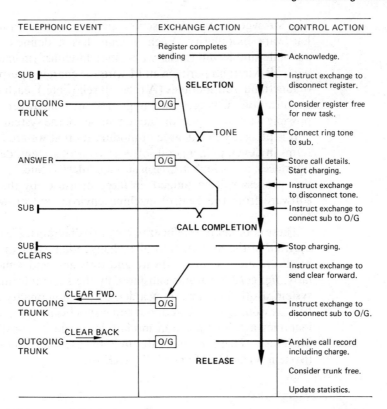

Call information	Charging determination storage Statistics keeping Performance monitoring
Exchange operation	Changing data (more generally, man-machine communications, MMC) Monitoring for faults
Signalling	Protocol conversion (to/from telephonic event into signal condition)

DATA TYPES

Permanent data	Exchange and network configuration
Semi-permanent data	Terminal information, i.e. subscriber directory number (DN), equipment number (EN), and class of service (COS).
Temporary data	Call data necessary per call

The Strowger system represents a perfect example of modularity, admittedly hardware modularity. Each module has a defined, limited purpose, has defined, simple interfaces with other modules (in most cases just *a* and *b* wires, P-wire and perhaps an M-wire for charging purposes), and is made up of functional sub-modules (A relay, B relay, etc.), each with its own functional specification. Changes and upgrades of a Strowger module are normally possible without impact on the remainder of the system.

It is precisely this degree of modularity that we are looking for in software terms in developing a good software control system: **defined purpose**, **defined interfaces**, and **a high degree of sub-modularity**. Success in this software definition process will be judged, in large measure, by the independence of the software from the control machine environment on which it is designed to run.

These comments are illustrated by two diagrams. *Figure 6.11* depicts a block diagram of the System X local exchange identifying the main modules, some hardware only, some hardware and software, and some software only. Similarly, *fig. 6.12* illustrates substantially the same information for the AXE 10 system. Both diagrams show a degree of commonality of the modules identified, as would be expected, but some significant differences. Neither diagram identifies a call connection module let alone its constituent parts identified previously. These modules are, in both cases, sub-modules of DSS and DSSS (System X) and LSM and GSS (AXE 10).

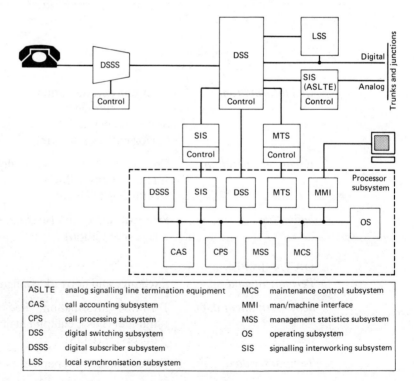

Fig. 6.11 Local exchange block diagram: System X

ASLTE	analog signalling line termination equipment	MCS	maintenance control subsystem
CAS	call accounting subsystem	MMI	man/machine interface
CPS	call processing subsystem	MSS	management statistics subsystem
DSS	digital switching subsystem	OS	operating subsystem
DSSS	digital subscriber subsystem	SIS	signalling interworking subsystem
LSS	local synchronisation subsystem		

Fig. 6.12 The subsystems used in a telephone exchange

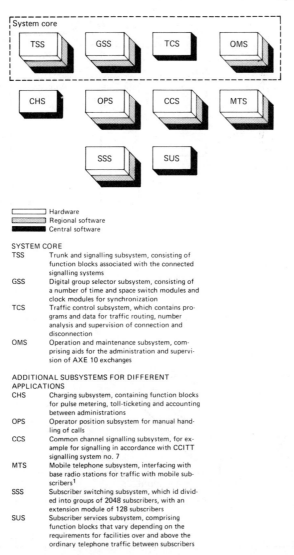

SYSTEM CORE
TSS Trunk and signalling subsystem, consisting of function blocks associated with the connected signalling systems
GSS Digital group selector subsystem, consisting of a number of time and space switch modules and clock modules for synchronization
TCS Traffic control subsystem, which contains programs and data for traffic routing, number analysis and supervision of connection and disconnection
OMS Operation and maintenance subsystem, comprising aids for the administration and supervision of AXE 10 exchanges

ADDITIONAL SUBSYSTEMS FOR DIFFERENT APPLICATIONS
CHS Charging subsystem, containing function blocks for pulse metering, toll-ticketing and accounting between administrations
OPS Operator position subsystem for manual handling of calls
CCS Common channel signalling subsystem, for example for signalling in accordance with CCITT signalling system no. 7
MTS Mobile telephone subsystem, interfacing with base radio stations for traffic with mobile subscribers[1]
SSS Subscriber switching subsystem, which id divided into groups of 2048 subscribers, with an extension module of 128 subscribers
SUS Subscriber services subsystem, comprising function blocks that vary depending on the requirements for facilities over and above the ordinary telephone traffic between subscribers

This discussion has treated modularity at the highest levels and *figs 6.11* and *6.12* are at a similar high level. The modularity must be extended downwards through multiple levels so that individual programs of, ideally, no more than 50 words build into sub-modules using the simplest of interfaces, the sub-modules into minor modules, and these, in turn, into modules all with equally simple and well-defined interfaces. An example of a possible submodule assembly of programs is given in *fig. 6.13*. Considering that the telephone exchange control machine is going to operate in real time upon random traffic it is only by means of rigid modularity in design that sanity of the system (and of the operating personnel) can be maintained.

Fig. 6.13 Microprogram execution: AXE 10

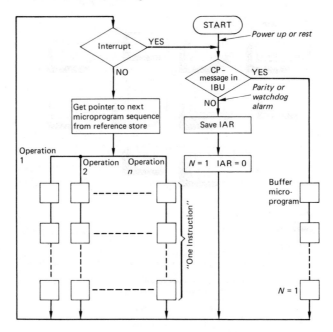

Fig. 6.14 Store allocation: AXE 10

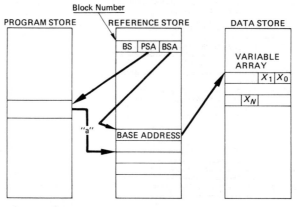

BS = Block state
PSA = Program start ADDRESS
BSA = Base start ADDRESS
"a" = Base address NUMBER (program constant contained in the current instruction)

Figure 6.14 returns to the other division, that between data and program, indicating how these are stored and caused to relate. This is the sort of diagram which, at first sight, has little to say, but further study reveals that it requires pages of explanation. *Very* simply, each block number refers out to a program and a base address. Each read or write instruction of the program contains a parameter *a*, the base address number which, with the base address, defines the area in the data store. For our purposes it is only necessary to appreciate that special attention has been paid to modularising software program and data.

Software Structure

The discussion of software has entirely ignored the nature of the control machine and taken no notice of the control architecture, duplicate synchronous, cold stand-by, etc. In treating the subject in this way we have simply been following the design objective of a good software structure.

What is "good" in this connection is perhaps difficult to fathom from the conflicting claims of competing manufacturers all of whom claim excellence of software structure. The objective is, however, simple and has been stated: "good" software is modular down to very small segments. The designer of a segment works within a confined set of interfaces and to a confined specification. Software changes either to rectify mistakes or to incorporate improvements will, if the design rules are successful, be similarly confined to one segment. To illustrate this concept we will use the software structure of ITT System 1240. Because of the distributed control nature of ITT 1240 a rigid modularity of software is of even greater importance. Modularity is achieved in several different, interlocking ways.

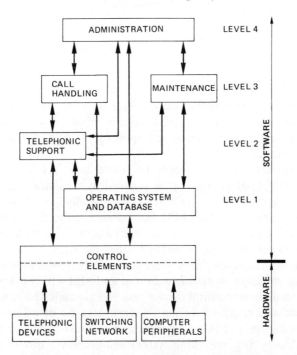

Fig. 6.15 Software structure: ITT 1240

Figure 6.15 illustrates the ITT 1240 concept of structuring software in levels. The lowest levels communicate with the *hardware* of the system via software modules called device handlers (not shown but discussed later in conjunction with *fig. 6.16*). It is only these lowest levels whose structure is influenced by the hardware environment. Higher levels are progressively more divorced from the hardware and intercommunicate via defined software message interfaces.

A second dividing concept is that of the *virtual machine* (*fig. 6.16*). The

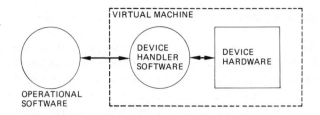

Fig. 6.16 Virtual machine concept: ITT 1240

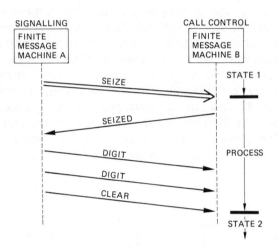

Fig. 6.17 Communication between finite message machines: ITT 1240

hardware and the software device handler, which alone has to be designed with the hardware characteristics in mind, are considered as a single "virtual machine" which is operated upon by the true software via standard software interfaces.

A third dividing concept illustrated in *fig. 6.17* is that of *functional division*. In ITT 1240 the concept is realised by a software structure based on "finite message machines" (FMM). This is really a method of very rigidly defining the interface. The FMM lies passive, expecting a particular message. On receiving that message it performs a series of defined actions which may result in its sending and receiving further messages. The FMM completes its task by arriving at another passive state where it awaits another particular message. The messages initiating action in an FMM will contain an amount of priming information required for the task and are called basic messages. The messages sent and received during the task may be of a less-detailed nature if they are between FMMs cooperating on this task and are called directed messages.

Every SPC switching system will employ a mix of such modular concepts to make the task of designing, de-bugging and maintaining the software possible. The language used to describe the structure varies and the success achieved is variable.

Operating Systems

We have discussed the provision of software structures, which make the software very modular, and software strategies and concepts, which divorce the software from the machine which it is designed to control. The eventual aim is to load the well-designed well-structured software on to the equally well-designed hardware, switch on, and see the new machine—a machine made of integrated hardware and software portions—work. We must be very clear about it. The new machine is complete only when software runs correctly on hardware and, for telephone exchanges, does so in real time for random traffic.

The portion of the design which provides the "personality" to the eventual hardware/software integrated machine is the operating system. The operating system will know the configuration of the hardware, how many CPUs, how much direct storage, how much back-up storage, input/output speeds, destinations, and protocols.

The operating system will also know where to find the initialising programs required to bring the system to a working state from cold. It will know how to disable parts of the system which no longer operate correctly. And it will retain very clear ideas of priority and will monitor work in progress to ensure that work of the correct priority is performed in the appropriate order.

Perhaps as much as the proper structuring of software, the operating system is crucial to the complete machine. To deal with the subject in less than a chapter is impossible; to devote a chapter to it is inappropriate. The reader is offered references and glimpses of the application of operating systems in the diagrams.

Figure 6.15 illustrated the position of the operating system in the machine hierarchy, very much the arbiter of the database and of the hardware configuration. *Figure 6.18* depicts the AXE 10 APZ 210 maintenance activities. Much of this survey treats of operating system elements. There is manual involvement by maintenance staff indicated by rhomboids. There will be operating system elements for functional entities within the overall control machine; one such, for a distributed control machine, is shown in *fig. 6.19*. Finally, *fig. 6.20* illustrates a process which could be part of a detailed operating system, started by interrupt and running through a number of extension modules (EM, AXE 10 terminology) performing tasks as necessary.

Software Development

The fact that a modern SPC telephone exchange has most of its logic processes expressed in software makes it necessary for the content and structure of the software to be known and controlled at all times.

A UK Strowger director depended for its translation information upon the position of wire jumpers in a number of translation fields. Once these jumpers were soldered in the correct position, the translation could be expected to be always performed correctly. In an SPC exchange, the translation is data, stored electrically and electrically alterable. The correct reading of the translation depends on the correctness of the data plus the correct operation of the software logic, itself stored electrically and electrically alterable.

We therefore perceive a need to keep back-up records of software programs

122 Introduction to Digital Communications Switching

Fig. 6.18 Maintenance activity system (MAS) function survey: AXE 10

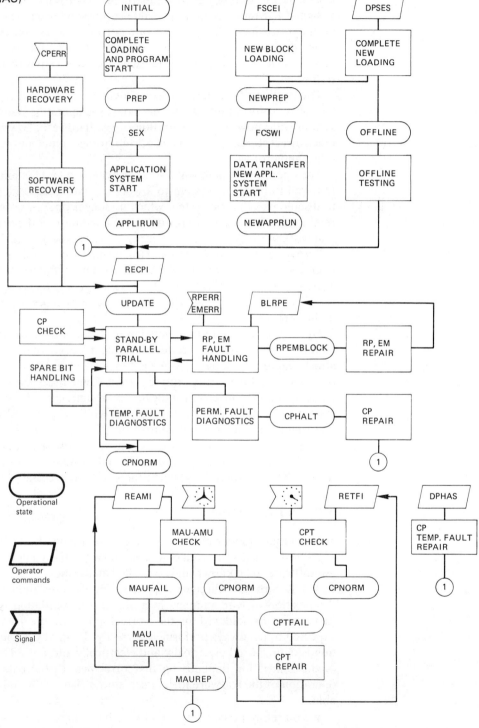

and data to protect against degradation of the working copies. There is a further need, or at least desire, to sub-divide software and data so that re-loads can be of modules only, not of the complete exchange. We can also perceive the need for the engineer to be able to read and understand the software logic and data in order to detect possible software sources of exchange malfunction. This too is an activity which is more easily performed in a modular software environment and requires that the software is expressed in a language easily comprehended by humans as well as by machines.

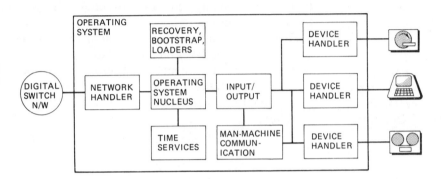

Fig. 6.19 Operating system as implemented in the control element of the computer peripherals module: ITT 1240

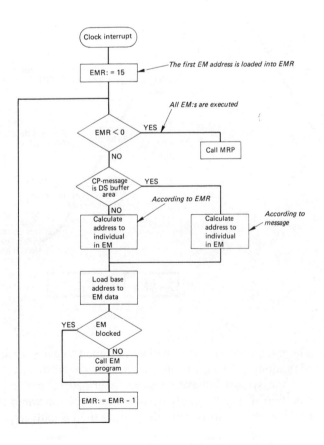

Fig. 6.20 Regional executive program: AXE 10

In apparent conflict with these requirements is the need to ensure that a particular exchange is equipped with software logic, data modules and hardware modules which are compatible in the sense that they will work together. For example, a software program to read from store a 3-digit translation is incompatible with a data storage plan that assumes 2-digit translations.

Thus the very improvements provided by modern systems—the ease of designing in software and designing in a modularised, structured way—bring their own problems. Ease of designing gives flexibility; flexibility allows frequent changes; frequency of change leads to problems of compatibility; maintaining compatibility requires a structured design process which leads back to ease of design.

Figure 6.21 demonstrates one manufacturer's view of the design process starting at the top level, applying testing to every piece of code written and repeatedly testing as coded programs grow to sub-modules, assemble into modules and thence into working systems. With remarkable infelicity of expression, the engineers concerned describe it as "top-down design, bottom-up testing".

Fig. 6.21 System X software development strategy (*POEEJ* VOL 72 Jan 1980)

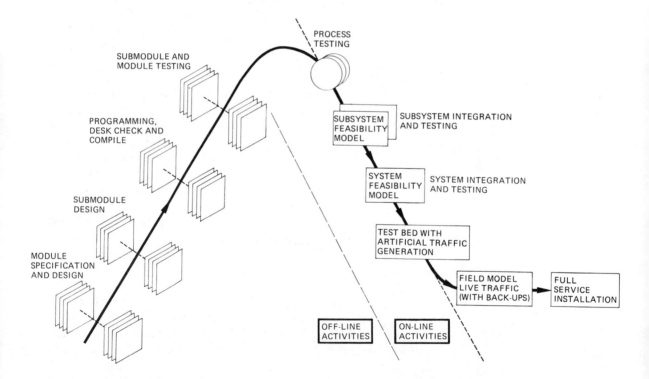

Overload Control

The telephone exchange will be designed to carry a specific load. The nature of random telephonic traffic dictates, however, that overload is always possible and the system behaviour under overload is very important. The overload situation will quite likely occur on an occasion when the continued capacity of the exchange to switch calls is more than usually important.

Fig. 6.22 Exchange overload characteristic

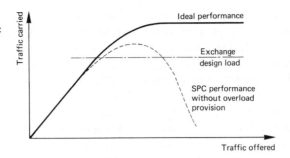

Figure 6.22 shows the performance of a typical exchange in carrying traffic under various applied loads of offered traffic. The ideal characteristic, achieved by the step-by-step system, is for the exchange to accept all offered traffic up to its design load and to continue to switch this maximum load under any level of (even extreme) overload in offered traffic.

Any system which, like Strowger, treats the complete call will approximate to this ideal characteristic. Systems using centralised control and, more particularly, systems which treat the call as a number of separate processes (pre-selection, selection, call completion, etc., for example) will run the risk of dealing under overload with portions of the process for a call which is destined not to be completed. The exchange control therefore becomes bogged down in unproductive activity as overload approaches and increases.

All SPC systems will experience this problem and the more modular the system the more it will be likely to engage in unproductive activity under excessive load. The self-regulatory mechanism of Strowger, that if a call will not be completed then processing of the call is not commenced, is absent in SPC systems. The SPC designs must, therefore, include specific overload detection and overload control features. These shall result in a smooth response to overload, gradually refusing new traffic as overload increases and gradually removing the inhibition on new traffic as the overload decreases. The success of such an approach depends on the accuracy and speed with which the onset of overload can be measured and controlled.

Chapter Summary

The introduction to the chapter highlights the very recent move away from a historical trend to combine increasing centralisation of control with increasing use of stored program control (SPC). Only in quite recent years has the advent of large-scale integrated devices encouraged the use of, or the claim to use, distributed control.

The chapter then dealt with this historical trend (towards centralisation) by treating step-by-step systems, i.e. distributed control, in some detail. This enabled the problems of network numbering and the use of translation to provide large local and national common numbering systems to be introduced. Link frame, marker-controlled systems received less-detailed treatment in following the historical trend towards centralisation.

A second approach to the problems of control was made via the various solutions to the problems of system security. This treatment introduced duplicate control, dual load sharing, multi-processing, and, full circle, distributed control.

The move towards centralised control was encouraged by the use of electrically alterable logic—software. The chapter introduced software concepts via a treatment of telephonic call progress. The need for modularity, already demonstrated for hardware, was shown also to be essential in software systems. The concepts of modularity and of the separation of the hardware and the software of the machine was used to discuss the proper structuring of software.

The chapter ended with topics also relevant to areas outside the field of telephone exchanges: an operating system knits a particular conglomeration of hardware modules, software programs and data sets into a very individual system. Lastly, the software development structure and cycle was shown to have a direct impact on the eventual maintainability of the system. Finally, there was a brief discussion of the problems posed by traffic overloads.

References

[6.1] P J Povey, *The Telephone and the Exchange* (Pitman 1979).
[6.2] A E Joel Jr (Ed), *Electronic Switching: Central Office Systems of the World* (IEEE Press 1976).
[6.3] M T Hills and S Kano, *Programming Electronic Switching Systems* (Peter Peregrinus 1976).
[6.4] *Electrical Communication*, Volume 56, No. 2/3, 1981: collection of articles on ITT System 12.
[6.5] *Ericsson Review 1979–82:* the review has published a collection of articles on AXE 10.
[6.6] *Post Office Electrical Engineers Journal* (now entitled *British Telecommunications Engineering*): a collection of articles on System X was published in 1981.

Exercises 6

6.1 List three ways in which the control task of a telephone exchange differs from the control task of a digital computer running pay-roll and stock-control systems.

6.2 Explain why the text mentions the H relay, used in Strowger selectors for switching, whereas *fig. 6.3* shows two such relays, HA and HB.

6.3 *Fig. 6.23* illustrates a local exchange network for a medium-sized town and its suburbs. Devise an area numbering scheme suitable for this network.

6.4 List the translation requirements that the numbering scheme you devised for Exercise 6.2 imposes on the exchanges of the network.

6.5 List the advantages and disadvantages of the form of distributed control employed in the Strowger step-by-step system.

6.6 Using *fig. 6.7* as a guide, list the functions performed by the following:
Markers
Originating registers
Incoming registers
Register access switch
Register junctor.

6.7 Produce a tabulation summarising the various advantages and disadvantages of the methods of control shown in *fig. 6.9*.

Fig. 6.23 Urban network

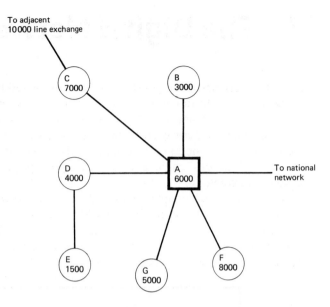

6.8 Suggest control architectures suitable for the following conditions:
 a) Calls in conversation phase on breakdown are retained, calls being established are lost.
 b) All calls, including those in process of being set-up, are to be retained on breakdown.

6.9 Produce a diagram rather more detailed than *fig. 6.9e* showing the location and communication arrangement of such necessary peripherals as man/machine communications, back-up program and data stores, etc.

6.10 Provide a list of the probable advantages of a distributed control architecture.

6.11 Now provide a list of the dangers and pitfalls to the designers and potential users of a distributed control system.

6.12 Produce a contrast list: the claimed advantages of SPC and the contrasting snags.

6.13 Provide examples of what is meant by "data" in telephone exchange control.

6.14 Define the qualities of a module which determine whether it is "hardware only", "mixed hardware/software", or "software only".

6.15 Given that software is, ideally, produced in several levels of abstraction from the hardware, write the level 1 to level 4 (level 1 nearest hardware) against the listed software functions without referring to *fig. 6.15*:
 Return of dial tone
 Operating system
 Determination of call charge
 Definition of number of digits required to route call
 System defence on fault occurrence
 Man/machine communication.

6.16 Explain the concept of a virtual machine.

6.17 Explain the problem of overload control in a central control SPC system.

7 The Digital Network

From the very first pages, this book has emphasised the primary importance of the network. In discussing communications switching we are dealing with a single world-wide machine. Without a doubt the present telephone network and the integrated network of the future represents the single largest and most complex engineering investment in the world. It is a satisfying reflection for communications' engineers that it also perhaps provides the greatest good of all our technology.

Having spent two chapters within the communications switching exchange we now look outwards to the network; a network still largely analog but with an increasing digital component. How is this digital component to be synchronised over areas, nations, continents and world-wide? How are these digital exchanges to converse with each other and with their subscribers?

Fig. 7.1

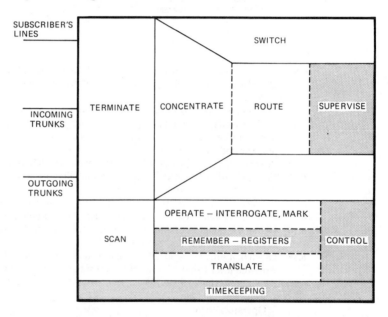

The subject matter of the chapter is illustrated in *fig. 7.1*. This time, however, the diagram which has served so well has had to be extended by introducing the concept of timekeeping which is strictly only relevant in systems using time division of switching or control. Also, the signalling functions are a part only of the supervision, register and control portions of the diagram.

Network Synchronisation

The introduction of time division either in control or in switching requires a timekeeping function as part of the exchange system. Time divided transmission links to other exchanges in the network introduces the further requirement that the exchange timekeeping be synchronised to the network.

Local Timekeeping—The Pulse Generator

Elementary pulse generators are present in all automatic systems. The Strowger system employed a rotating generator to generate tones and ringing cadences, and cam-operated contacts driven by the generator were used to provide the exchange with timing pulses for alarm delay timing and other purposes. The exchange also had an accurate pendulum clock to provide real time indications to, for example, change from day-time to night-time charge rates.

The introduction of time divided common control increased the need for timing and a pulse generator system was introduced to supply, as a basic minimum, an accurate pulse supply at the highest frequency used by the time divided control. Such supplies could be, and sometimes were, synchronised to real time so that the SPC control could provide dated and timed fault reports and other man/machine communication.

When time divided switching and transmission is introduced, the pulse generator must be extended in range to supply a timing signal at the basic PCM rate (2.048 Mbit/sec for CEPT 30-channel) and a sub-set of this fundamental frequency will time the frame and multi-frame and, also, define the clock frequency required by control.

Summarising the different timekeeping requirements:

Control machine clock frequency.
Scanning frequencies for telephonic devices.
Timing of PCM stream, frame and multi-frame.
Timing of delays and periodic happenings.
Provision of calendar and time-of-day information.

In a practical exchange system the first three will probably be provided by a pulse generator system whereas the last two will be provided by a separate real time calendar clock. The pulse generator will be common to the exchange and absolutely essential to its operation. It is, therefore, subject to the very highest security requirements.

What is not evident from the previous discussion is that the pulse generator forms a very separate sub-system of a kind that is quite unlike the exchange system and very separate from it. Pulse generators and power supplies are both similar in this respect. Both are essential to the exchange system but both may use design techniques very different to those of the exchange system.

Security of the pulse generator is assured most easily and effectively by triplication, which allows simple majority decision logic to supply the exchange with good timing from the remaining good pulse generators. Part of the pulse generator, however, is the exchange distribution wiring which must also be secured. This can be achieved by duplication but then there is the problem of identifying the bad distribution. One solution to these problems, utilised in the UK TXE 4 system, was to provide four pulse generators, one slaved from the triplicated three, and two distributions. This combination allowed the bad distribution to be disabled by the pulse generator error check comparing the duplicate return with the quadruple output.

Network Timekeeping—Synchronisation

The need for network synchronisation is, as noted, a new requirement introduced to communications switching by the use of PCM TDM switching and transmission. It is only possible to transmit coded time-divided information end to end through a switched network provided every node of the network can recognise the individual elements of the message. This can only be effected by keeping the network synchronised.

Methods of maintaining synchronism are summarised in *fig. 7.2*. One is to provide highly accurate clocks at each exchange, so accurate that there is almost absolute coincidence in timekeeping, and so stable that drift of phase difference between the many clocks is avoided. This is known as *plesiochronous operation*. Caesium standard atomic clocks have the necessary qualities of accuracy and stability but are expensive and of indifferent reliability.

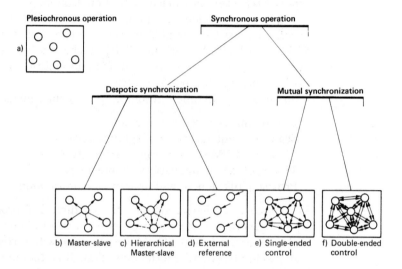

Fig. 7.2 Network synchronisation concepts

Alternatively, there are several forms of synchronous operation also illustrated in *fig. 7.2*. The alternatives shown are ordered *b* to *f* in steps of probable increasing complexity but also probable increasing reliability. The mutual synchronisation methods also tend to overcome the effect of difference in ambient temperature which may cause phase differences in the despotic methods.

All the synchronous methods shown are realised by using a local oscillator which is phase-locked to the timing inherent in the bit stream of one of the PCM systems incoming from the master timekeeper. Temperature variations and variations due to aging cause the local oscillator to change its frequency and, therefore, phase differences are introduced. The choice of a synchronising method is, of course, dependent on the nature of the requirement for absolute synchronism. *Figure 4.24* illustrated one aspect of **slip**, the major effect of lack of synchronism. *Figure 7.3* indicates another view of the effect, this time particularly on a facsimile transmission.

Fig. 7.3 Example of the effect of slip on a facsimile signal

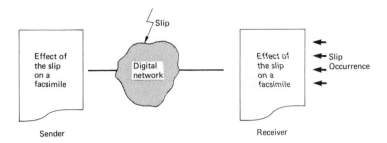

Submissions to the CCITT indicated that for voice communication a slip rate of 300 slips/hour would be acceptable. Slip results in a speech sample of 8 bits being lost or repeated on all 30 channels of a system simultaneously. This could be perceived as an audible click on about 1 occasion in 25. Assuming a connection through 8 exchanges and a timing variation of ±3 parts in 10^7, the average frequency difference may be two-thirds of the maximum variation. The frame slip rate in these conditions would therefore be

$$7 \times \frac{2}{3} \times \frac{3 \times 10^{-7} \times 2.048 \times 10^6 \times 3600}{256} \simeq 40 \text{ slips/hour}.$$

Should only 4% of these be noticed then the noticeable slip rate for voice communications is 1–2 slips per hour.

The effect of slip on communications signalling will depend on the type of signalling. Voice frequency signals encoded in the PCM stream will be subject to a phase discontinuity which is of no significance to the detectors. Channel associated signalling in channel 16 on the other hand may suffer from loss of multi-frame alignment due to slip. Alignment may take up to 5 msec to be regained and calls in process of being connected may be lost. Common channel signalling is equipped with error-detect and re-transmit features so that a slip will cause increased re-transmit activity but the signalling will not be otherwise affected.

Data transmissions will be affected, the severity being dependent upon the error-handling features incorporated in the transmission protocol. A performance criterion used is the percentage of error-free one-second intervals (EFS). A figure of 99.5% EFS as an operational objective incorporating errors from all causes is desirable and, if errors due to slip only are considered, then a target of 99.9% EFS is thought to be appropriate. In the example quoted above, 40 slips per hour means 40 one-second intervals containing errors in the hour. Hence

$$\text{EFS} = \frac{3560}{3600} \times 100 = 98.8\%.$$

The slip rates achievable with plesiochronous operation are, for example, shown below for a clock of caesium-type accuracy and one of the accuracy of, say, a quartz oscillator. It is assumed that there is an 8-bit buffer (*fig. 4.23*).

Fig. 7.4 Cost versus long-term stability for a single clock unit

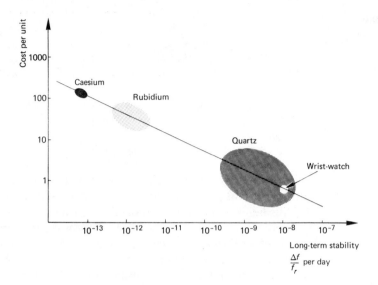

	Pulse rate bits/sec	Frequency accuracy	Bits accumulated per day	Slips per day
Quartz	2048.10^6	10^{-7}	17 695	2212
Caesium	2048.10^6	10^{-12}	0.18	0.02

Figure 7.4 illustrates the cost differential between increasingly accurate clock systems. The caesium standard clock, though accurate, is expensive and not remarkably reliable. To maintain such a clock as a reliable timekeeper would require at least triplication, and units requiring repair would probably need to be returned to the manufacturer because of the rather special nature of the equipment.

Synchronisation Strategies

All that has been said indicates that a synchronisation strategy will probably consist of an amalgam of several of the methods so far described. CCITT requirements are most easily met by national administrations maintaining plesiochronous synchronism on international links. This suggests a national reference that is of caesium accuracy. Within the national network some sort of hierarchy is indicated, slaved probably to the national reference but using a mixture of master/slave and mutual synchronism. Such a mixed arrangement was illustrated in *fig. 4.22*.

Synchronisation Systems

Figure 7.5 illustrates the system requirement for a mutual synchronisation system using single-ended control. The frequency of the incoming signal is compared with the local clock frequency and the difference is obtained from the elastic buffer. This difference represents the average of the exchange clocks and is used via the compensator to adjust the frequency of the local clock.

The Digital Network 133

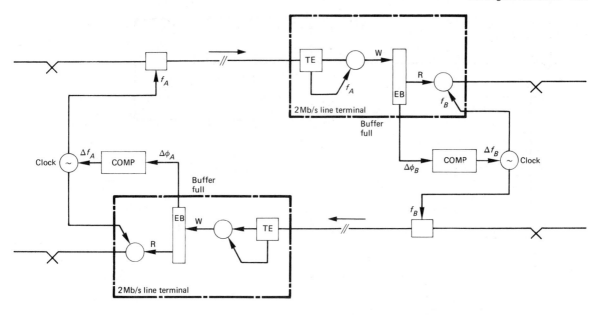

Fig. 7.5 Mutual synchronisation: single-ended control

TE	Timing extract
COMP	Compensator
EB	Elastic buffer
R	Read
W	Write
f_A, f_B	Clock frequencies at exchanges A, B
$\Delta\phi_A$, $\Delta\phi_B$	Instantaneous phase difference measured at exchanges A, B
$\Delta\psi_A$, $\Delta\psi_B$	Instantaneous corrected phase difference calculated at exchanges A, B
Δf_A, Δf_B	Frequency correction applied to clocks at A, B

Simultaneous with adjustment of exchange clock B by clock A in this manner, exchange clock A is being adjusted by clock B.

Single-ended control compensates for the gradual drift of the clocks but fails to compensate for changes in ambient temperature. *Figure 7.6* illustrates the more complex double-ended control which compares the difference signal obtained at both ends of the synchronising link and applies the resulting corrected difference as a clock adjustment. This added sensitivity allows compensation for frequency changes due to temperature variation and also lessens the effect on the network of inserting a new clock running at a frequency remote (within the limits allowed) from the system frequency. Such an introduction, either through re-adjustment of a clock or introduction of a new exchange, will cause a ripple effect on a mutually synchronised network as all clocks adjust to agree with the new member. This ripple will be much less marked where double-ended control is in use.

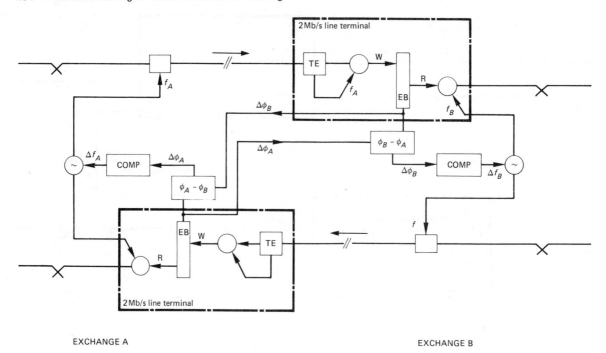

Fig. 7.6 Mutual synchronisation: double-ended control

A despotic control system would consist of just half of *fig. 7.5*.

Figure 7.6 indicates a need to transmit information on the phase difference determined at exchange B back to exchange A and vice versa. This information is precisely the kind for which the spare capacity of channel 0 was intended. *Figure 7.7* illustrates the allocation in channel 0 defined in the UK system for this purpose.

Signalling

Background

The signalling function has not been discussed since it was introduced as a concept in Chapter 1. To re-introduce it we will return to the original manual operator controlled telephone networks.

The earliest operator boards were used in conjunction with local batteries connected to each telephone and providing the DC current feed to the microphone. The subscriber called the exchange by turning the handle of a magneto generator which caused the "eyelid" of an eyeball indicator in the manual board to open. This audible noise and visual indication of the calling line indicated the calling signal, one of the *line signals*.

The *routing information* (signals) was given to the operator verbally by the caller and the operator could then complete the call by plugging in to the wanted line or to a trunk to another exchange. In either case the trunk or line could be seen to be busy because its indicator was uncovered, the *busy signal*. In early systems a release signal was difficult to arrange and the operator had to monitor the call looking for a clear.

Fig. 7.7 UK CEPT 30-channel: channel 0 bit 5 allocation for synchronisation

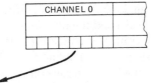

Bit 5 contains 32-bit word repeated twice every 16 msec, i.e. one bit is sent every second channel 0.

BIT		
0	0	
1	P	Parity of bits 2 − 9
2	⎫	
3	⎬	Reset command
4	⎭	
5	⎫	Oscillator control
6	⎭	
7	⎫	
8	⎬	Negative acknowledgement
9	⎭	
10	0	
11	P	Parity of bits 12 − 19
12	⎫	
13		
14		
15	⎬	Δφ
16		
17		
18		
19	⎭	
20	0	
21	1	
22	1	
23	1	
24	1	
25	1	
26	1	
27	1	
28	1	
29	1	
30	1	
31	1	

An early major advance in telephone switching was the introduction of the central battery system when indicators were replaced by lights, loop signalling replaced the magneto, busy was indicated by touching the tip of the plug to the sleeve of the jack, and release was indicated by re-lighting the calling lamp.

Analog Signalling Methods

Strowger's invention of the dial introduced automatic routing by loop-disconnect signalling and this remained as the industry standard for 70 years and will be a component of the network for some time to come. Line signalling, on the other hand, was forced to change to AC voice frequency messages because of the introduction of amplifying repeaters and FDM carrier telephony.

As operator dialling of international calls and, more recently, subscriber dialling of national and international calls, was developed, the need for more sophisticated signalling systems increased. All such signalling systems were channel associated and it was not until the advent of SPC exchanges that any other method of signalling was even considered. Stored program control of individual exchanges revealed the possibility of control machines cooperating directly in network control architectures. If the control machines themselves are to converse then it is a small additional conceptual step to consider common channel signalling between the control machines carrying the signalling information for many hundreds of speech connections.

One illustration of the economic gains possible with common channel signalling is the continental operator dialling signalling relay sets introduced in the late 1950s on circuits from London. On each circuit the terminating relay set contained some 30 relays, all but one or two associated with the two-voice frequency signalling system. Common channel signalling introduces a vastly more complex system sharing the signalling task of hundreds or thousands of individual voice circuits.

It is not intended to elaborate on existing analog signalling techniques; several example systems are illustrated in *fig. 1.12* and *figs 7.8–7.12* and the information contained in these diagrams is sufficient to prepare for the later discussion of analog to digital signal interchange.

The various categories of signalling system are summarised in the following:

DC Signalling

LOOP DISCONNECT   Line and selection signalling. Limited by line length as pulse distortion increases due to cable capacitance.

LONG-DISTANCE DC   Based on the use of symmetrical waveforms. Used for line and selection signals. Increases signalling range by a factor of 3 or 4 over that obtained by loop disconnect.

VF AC Signalling

OUTBAND   Signals sent within transmission frequency band but outside voice frequency band, typically 3825 Hz. Simple system with small signal repertoire dependent on transmission media of full 4 kHz range.

INBAND   Signals sent within speech band. Danger of voice imitation limited by use of guard circuits to ensure that signal frequency is not accompanied by other frequencies as would happen with voice imitation. Line splitting is required to prevent signals being transmitted onward beyond the node at which they are directed.

LINE SIGNALLING   Both outband and inband 1VF systems are used as the line signalling system in conjunction with another more sophisticated system for selection signals.

MULTI-FREQUENCY SYSTEMS   Inband multi-frequency systems, usually 2 frequencies out of n (n being either 5 or 6). Much less susceptible to voice imitation and, of course, with a much greater signal repertoire. Used as selection (register) signalling systems.

Fig. 7.8 General arrangement for VF line signalling

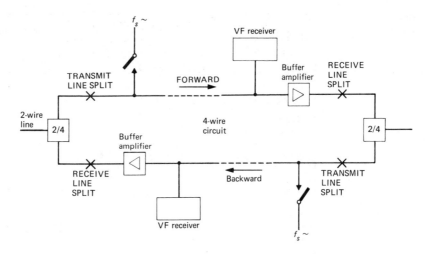

SPLIT — Line splitting prevents onward transmission of signals.

BUFFER AMPLIFIER — Prevents signals from the sending end leaking across the 2/4 wire bridge affecting the VF receiver at the sending end.

Fig. 7.9 Tone-on idle continuous VF signalling: Bell SF signalling

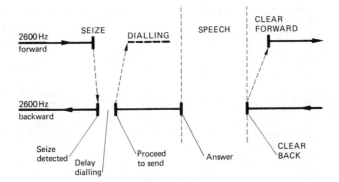

Digital Signalling

In the four practical PCM systems described in Chapter 3 arrangements were included to accommodate a digital version of the major analog signalling methods. This was particularly important for the US systems where outband signalling was used widely and a digital "outband" was provided per channel using bit stealing techniques. Inband VF signalling can of course be passed transparently over the digital channel and there is, therefore, no pressing need to provide a digital version for, say, the UK SSAC9 1VF system (*fig. 7.10a*). The UK, however, also uses DC signalling techniques extensively, particularly on local trunks. (In the UK a local trunk is called a junction and we talk of the local junction network.) Since the digital transmission systems were introduced to the UK primarily for the local junction network then, not surprisingly, the UK 24-channel system and the CEPT 30-channel system were equipped with quite sophisticated channel associated signalling arrangements.

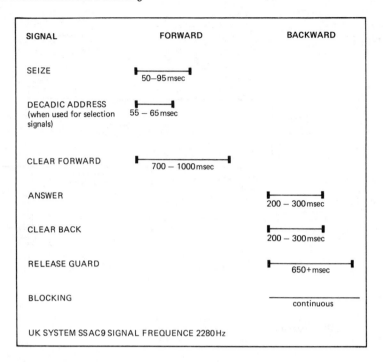

Fig. 7.10a Pulse VF signalling

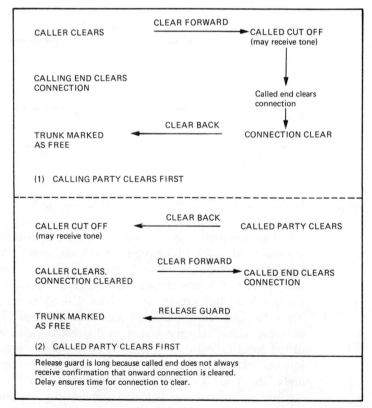

Fig. 7.10b Clear signal protocol

Fig. 7.11 Push-button telephone signalling: UK system SSMF 4

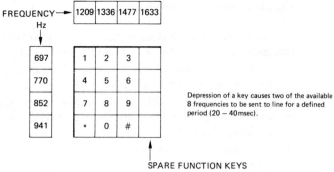

Depression of a key causes two of the available 8 frequencies to be sent to line for a defined period (20 – 40 msec).

SPARE FUNCTION KEYS

Fig. 7.12 International CCITT R2 signalling MF

TWO OUT OF SIX SIGNALLING, SIGNAL FREQUENCIES.

FORWARD	1380	1500	1620	1740	1860	1980 Hz
BACKWARD	1140	1020	900	780	660	540 Hz

FORWARD SIGNALS   DIGITS 1 – 0.

CODE 11 OPERATOR.

CODE 12 OPERATOR.

CODE 13 AUTOMATIC TEST EQUIPMENT.

COUNTRY CODE

ECHO SUPPRESSOR CONTROL.

ORIGIN OF CALL IDENTITY.

LANGUAGE DIGIT.

CLASS OF SERVICE.

END OF PULSING.

FORWARD TRANSFER.

SPARES (NATIONALLY ALLOCATED).

BACKWARD SIGNALS   REQUEST ADDRESS TRANSMISSION.

SEND NEXT DIGIT, REPEAT LAST BUT ONE DIGIT, etc.

REQUEST CLASS OF SERVICE.

INTERNATIONAL/NATIONAL CONGESTION.

NUMBER COMPLETE.

CALLED LINE CONDITION.

CHANGED NUMBER.

SPARE (NATIONALLY ALLOCATED).

Reference
CCITT RECOMMENDATIONS
Q400 to Q480

Channel Associated Signalling

It is only intended to discuss the signalling arrangements of the CEPT 30-channel system (refer to *fig. 3.13* and Table 3.6 on page 45).

The 4-bit quartet repeated once every 16 frames provides a signalling capacity per channel of 2 kbit/sec. This is ample capacity for any existing signalling system. The ready availability of this signalling capacity has led to the adoption of digital versions of some of the analog systems, notably CCITT R2 (Recommendations 421–424). This digital version is for the line signalling part of the R2 system only. Use of a digital line signalling system simplifies the digital line termination equipment as it does not have to decode line signals contained in the bit stream. *Figure 7.13* illustrates this in depicting the line signal extraction arrangements at the exchange termination of a 2.048 Mbit/sec digital link.

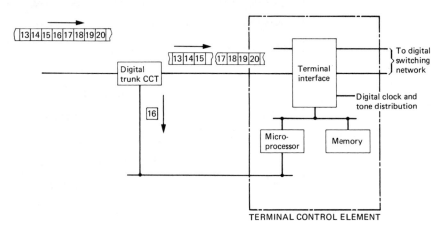

Fig. 7.13 Channel 16 signalling extraction: ITT System 1240 digital trunk module

Channel associated signalling is extracted from the incoming bit stream at the digital line termination (*fig. 7.13*) and thence the signals are used to effect the requisite control actions. Thus, beyond the digital line termination the functions of channel 0 and 16 disappear and, through the switch, they can be used as normal channels carrying voice traffic. In the case illustrated in *fig. 7.13*, that of ITT System 1240, the quartet information is used by the microprocessor in the terminal control element and data required elsewhere in the exchange is packaged into a message and launched over the digital network addressed to the target processor.

Channel associated PCM signalling is a natural choice where the PCM system is used just to replace an analog transmission system. Like all channel associated signalling, there is no need to "prove" the speech circuit: if it can't be signalled over, it can't be used. Normally the channel associated signalling will be used for line signalling only, although DC selection signals will probably also be converted into channel 16 signalling. Any VF selection signals, either push-button MF dial signals or multi-frequency inter-register signalling (such as CCITT R2), will be encoded as for voice and extracted by digital receivers in the registers. There is, therefore, no improvement in signalling speed when using channel associated signalling on PCM even though the SPC exchange controls are capable of working at much higher speeds than those dictated by conventional selection signalling systems. The slow speed of conventional selection signalling is perceived by the user as excessive post-dialling delay. This is made more obvious by the introduction of push-button dialling, whereas it was previously disguised by the slow speed of the rotary dial.

In summary then, conventional channel associated signalling methods, whether analog or digital, suffer from the following disadvantages:

- Signalling is relatively slow.
- Information capacity is limited.
- There is therefore a limited capacity to convey information which is not call-related.
- Some systems cannot signal during speech.
- The signal repertoire is difficult to change.
- Signalling per channel is, by its nature, expensive as the opportunity to share resources between channels cannot be realised.

Common Channel Signalling

The introduction of stored program control with very fast processors controlling the operation of the exchange immediately suggested direct cooperation between processors in different exchanges using high-speed signalling links. Quite apart from the signalling requirement per call connection, communicating processors could, it was argued, cooperate in much wider areas including charge determination and analysis, network routing and re-routing, and network maintenance.

High-speed processor control, particularly of electronic switches, also helps to reduce the post-dialling delay particularly when allied to the use of MF push-button signalling.

Cooperation between processors could also, it was argued, extend to the stage where not every exchange was equipped with the full processing power but was "supervised" by the processor in another exchange. Again this would only be feasible with high-capacity inter-exchange signalling.

These arguments led to early proposals (around 1966) to introduce the entirely new concept of **common channel signalling**. The concept is illustrated in *fig. 7.14* and is, in essence, the provision of a separate signalling channel, processor to processor, over which signalling data is carried for each of a (large) number of traffic circuits. In addition, the signalling associated with a particular circuit will need to be identified by including a label in the signal format. Clearly too, all the arguments about security that were reviewed in Chapter 6 in relation to exchange control apply to CCS also.

It is appropriate to summarise the advantages and disadvantages of CCS The latter, particularly, will then be given prominence in our discussion of CCS systems.

Fig. 7.14 Principle of common channel signalling

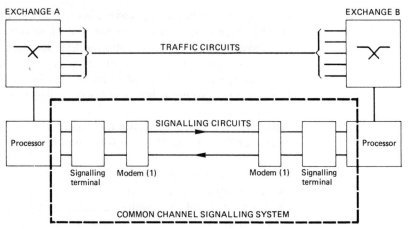

(1): in digital CCS the modem becomes a primary multiplex

Advantages of CCS
- Signalling is made entirely separate from switching and from traffic circuits and may therefore evolve without constraint from these functions.
- Signalling speed is increased, matching control processor speeds.
- Potential for a large repertoire of signals.
- Signalling possible at all times.
- Signalling protocols are flexible; signals can be changed or added.
- Signalling can include processor functions, network management, etc.
- Assuming that the traffic circuits served by CCS are numerous, then signalling is performed economically.
- For smaller groups of traffic circuits then a separate channel that is really separate (via different exchange nodes) can still provide economic signalling.
- No impact on the traffic circuits; functions such as line splitting, for example, are not required.

Disadvantages of CCS
(Disadvantage is too strong a word. There follows a list of requirements necessitated by CCS which were not present with channel associated signalling.)
- Security: errors in signal transmission will be intolerable if they are more than a low minimum. Error detection and correction are therefore required.
- Security: loss of the signalling link must not mean loss of all the many associated traffic circuits. There must therefore be adequate provision for back-up signalling links.
- Because signalling is over a separate channel there must be a separate function to ensure that the traffic circuit when connected is good.

Security of the Common Channel

The first internationally specified CCS system, CCITT signalling System No. 6 (#6 for short), had a capacity to deal with the signalling requirements of from 1500 to 2000 traffic circuits. (As an exercise, calculate the number of subscribers generating $0.04\,E$ each utilising such circuits at $0.9\,E$ loading who might be inconvenienced by loss of the CCS link carrying signalling information for all these circuits.) Clearly, security provision of the same order as that for a small telephone exchange is indicated. Three forms of CCS link architecture have been identified and given names. These are listed below and illustrated in *fig. 7.15*.

Fig. 7.15 Modes of common channel signalling

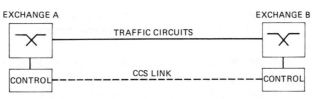

(a) Associated signalling exchanges A to B

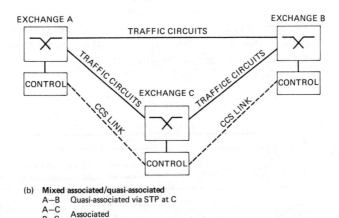

(b) Mixed associated/quasi-associated
A—B Quasi-associated via STP at C
A—C Associated
B—C

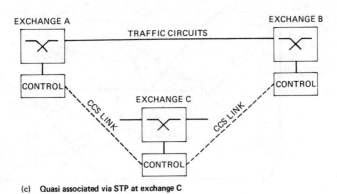

(c) Quasi associated via STP at exchange C

Associated CCS Signalling is transferred over a CCS link which follows the same route as the traffic circuits and terminates only at the exchanges between which the traffic circuits are connected. The arrangements shown in *fig. 5.27* between the concentrator and the parent exchange are associated CCS where channel 16 becomes the CCS bearer channel.

Non-associated CCS The signalling link follows a different route from the traffic circuits, passing via one or more transit exchange points. These exchanges, transiting signals but not necessarily traffic, are known as *signal transfer points* (STP). Non-associated signalling can be either:

Fully dissociated The signals are transferred over any available path in the network according to a set of comprehensive routing principles for the network.

Quasi-associated The signals are transferred over two or more links in tandem but only according to a predetermined routing.

Administrations will use a security scheme incorporating some or all of these modes to provide a secure CCS signalling architecture. In the UK a fully associated system is used. Each CCS link is provided in duplicate at least and up to a maximum of four links between terminating nodes. The CCS links are, for preference, given different routings between the nodes. Failure of the link causes the signalling traffic to be diverted on to the second link and further failures to the third and fourth if necessary. No diversion is allowed outside this group. Associated signalling is therefore retained even under the worst failure condition contemplated.

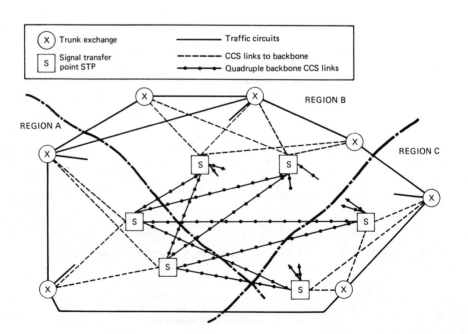

Fig. 7.16 Bell US quasi-associated CCS architecture

The US arrangement (*fig. 7.16*) is more complex as befits a larger network using a low-capacity system. The Bell system uses the 193rd bit of every second channel (reserved for multi-frame alignment in channel associated signalling applications) as the common channel and has therefore only 4 kbit/sec available for CCS compared with 64 kbit/sec on a CEPT system using channel 16. Partly because of this limited capacity the Bell system has standardised on #6 signalling. Direct links using associated signalling are provided from every exchange with CCS to both the signal transfer points in each region. The STPs of each region are interconnected with the STPs of other regions by a quadruple link arrangement with fall-back from link to link on failure. Direct associated links between trunk exchanges in different regions are provided as necessary in addition to the quadruple "backbone". The multiple fall-back options have led to the definition of a larger address field than agreed internationally by CCITT for the Bell version of #6.

Common Channel Signalling Systems

The CCITT has defined two systems for common channel signalling for international and national use. CCITT signalling system number 6 was defined in 1977 and intended for use as a high-speed digital system between processors using analog FDM links. It is based on a fixed word length of 28 bits, the word being called a **signal unit**, SU, and signal units are sent in blocks of 12. Error control is by retransmission of those SUs negatively acknowledged as having been received with errors.

This form of error control involves the possibility that SUs will eventually be received in the wrong order and that uncorrupted SUs may sometimes be retransmitted. These features impose an additional load on the receiving processor. Allied to this, the 28-bit SU format is not ideal for digital systems based on an octal system. For these reasons, work continued at the CCITT and a more sophisticated system specifically for digital use, the #7 system, was defined in 1980.

The UK development of System X proceeded in parallel with CCITT deliberations on #7 signalling. Courageously, perhaps, a fundamental concept of System X was to be the use of CCS between exchange sub-systems as well as in the network. There was therefore an urgent need for the UK to define a CCS system in advance of the #7 definition. The result is that the UK CCS called Message Transmission System, MTS, is similar to the #7 but by no means identical.

The UK has continued on this esoteric route by defining other #7-like systems for use on links to private exchanges, Digital Access Signalling System, DASS, and on private links between private exchanges, Digital Private Network Signalling System, DPNSS. The need for this clutch of new systems in addition to #7 is not self-evident. British Telecom claims that it would be unacceptable for customers as well as the administration to be allowed to use #7 signalling (or MTS) facilities.

In the following sections brief details will be included of #6 and #7 signalling only. This will be sufficient to introduce the concepts of CCS.

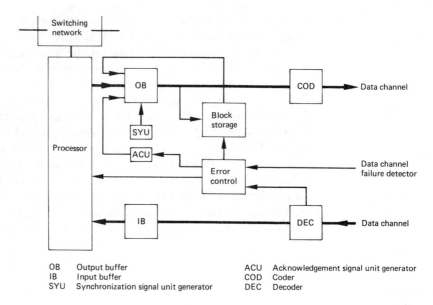

Fig. 7.17 Functional block diagram of a System No. 6 terminal

OB   Output buffer
IB   Input buffer
SYU Synchronization signal unit generator
ACU Acknowledgement signal unit generator
COD Coder
DEC Decoder

CCITT Signalling System Number 6

The System The #6 signalling system employs a basic word length of 28 bits. The word is called a signal unit and signal units are transmitted in blocks of 12, the 12th signal unit of each block being an acknowledgement signal unit.

Figure 7.17, reproduced from CCITT Recommendation Q251, illustrates the basic system concept of a #6 signalling terminal, while *fig. 7.18*, reproduced from the same source, illustrates the concepts and terminology of the constituent parts of the CCS signalling link.

The signals are formatted in the exchange processor and delivered to an output buffer in parallel form (on 28 signal leads). The buffer delivers the highest priority signal, or series of SUs constituting the highest priority message, to a coder in serial form. Each SU is encoded by addition of check bits in accordance with the check bit polynomial in use. The signal is then transmitted over the link in serial form. At the receiver end, the serial data is delivered to a decoder where each SU is checked for error by parity check of the check bits. Corrupted SUs are discarded and error-free message SUs are transferred to an input buffer after deletion of the check bits. The input buffer delivers the SUs to the processor in parallel form for appropriate action.

Signal Unit Format Signal units and the messages assembled from SUs are of various kinds as follows:

LSU   *Lone signal unit.* All the message information is contained in the one SU (*fig. 7.19a*).

The Digital Network 147

Fig. 7.18 Concept of a System No. 6 signalling link

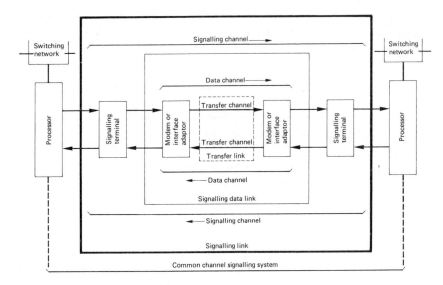

	Analogue version	Digital version
Data transceiver	MODEM	INTERFACE ADAPTOR
Transfer channel	(Voice-frequency channel) A one-way voice-frequency transmission path from the output of a data modulator to the input of a data demodulator, made up of one or more voice-frequency channels in tandem.	(Digital channel) A one-way digital transmission path from the output of the interface adaptor to the input of the interface adaptor, made up of one or more digital channels in tandem.
Transfer link	(Voice-frequency link) A two-way voice-frequency transmission path between two data modems, made up of one voice-frequency channel in each direction.	(Digital link) A two-way digital transmission path between two interface adaptors, made up of one digital channel in each direction.
Data channel	A one-way data transmission between two points, made up of a modulator, a voice-frequency channel and a demodulator.	A one-way data transmission path between two points, made up of a digital channel terminating on an interface adaptor at each end.
Signalling data link	A two-way data transmission path between two points, made up of one data channel in each direction.	
Signalling channel	A one-way signalling path from the processor of one switching machine to the processor of another switching machine.	
Signalling link	A two-way direction signalling path from processor to processor made up of one signalling channel in each direction.	

148 Introduction to Digital Communications Switching

Fig. 7.19 System No. 6 message formats

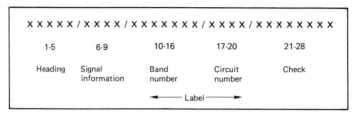

(a) Lone Signal Unit (LSU)
Initial Signal Unit (ISU) of multi-unit message (MUM)

```
0 0 / X X / X X X X X X X X X X X X X X X X / X X X X X X X X
1-2   2-3              5-20                       21-28
 *     **         Signal information              Check
   * Subsequent signal unit heading code
   ** Length indicator
```

(b) Subsequent signal unit (SSU) of MUM

MUM   *Multi-unit message.* A message consisting of a number of signal units. The complete message consists of
    ISU   Initial signal unit (*fig. 7.19a*) of a MUM followed by one or more:
    SSU   Subsequent signal units of a MUM (*fig. 7.19b*).

IAM   *Initial address message* (*fig. 7.19c*). At the start of every call connection set-up, an IAM is sent containing full address information. Subsequent messages relating to the same call can consequently be abridged. This is effectively the seizure signal. *Figure 7.19c* is an example, not only of an IAM but also of a 3-unit MUM.

ASU   *Acknowledgment signal unit* (*fig. 7.19d*). Sent as the 12th signal of each block, this responds to the sending end with an indication of the quality of the signal units of a defined block received. An acknowledgement indicator at 0 indicates an error and re-transmission is required.

SYU   *Synchronisation signal unit* (*fig. 7.7.19e*). Sent whenever there is no message signal, the SYU serves to maintain both ends of the link in synchronism as well as to function as a "filler".

The fields of the signal unit consist of:

HEADER   Indicating the type of signal, group of signals or message. For example, header 10 000 indicates the ISU of an IAM.

SIGNAL INFORMATION   This is the real content of the message.

LABEL   Indicating the circuit to which the message refers. The label is subdivided into band and circuit number fields so that the signal transfer point processor need only operate upon the band number.

CHECK   The series of check digits added as a result of inspecting the body of the SU and used at the receiving end to detect errors by parity check.

```
ISU      1 0 0 0 0 / 0 0 0 0 / X X X X X X X X X X X X / X X X X X X X X

         Heading  Signal              Label                    Check
         code     information
                  code


1st SSU  0 0 / 0 1 / X X X 0 0 0 0 0 X X X X 0 0 0 0 / X X X X X X X X

          *   **       Other routing information              Check


2nd SSU  0 0 / 0 1 / X X X X  X X X X  X X X X  X X X X / X X X X X X X X

          *   **      1st      2nd      3rd      4th           Check

                              Address signals

  *  Subsequent signal unit heading code
  ** Length indicator
```

(c) **Three-unit initial address message (IAM)**

```
         0 1 1 / X X X X X X X X X X X / X X X / X X X / X X X X X X X X

           1-3        4-14              15-17    18-20      21-28

         Heading   Acknowledgement        *        **       Check
         code      indicators

          * Sequence number of block being acknowledged
         ** Sequence number of block completed by this ACU
```

d) **Acknowledgement signal unit (ASU)**

```
         1 1 1 0 1 / 1 1 0 1 / 1 1 0 0 0 1 1 / X X X X / X X X X X X X X

                    1-16                        17-20        21-28

                Synchronization pattern           *          Check

          * Sequence number of signal unit in the block
```

e) **Synchronisation signal unit (SYU)**

Error Control The long-term mean bit error rate on analog circuits is about 1 in 10^5; on digital circuits it is about 1 in 10^7. Short burst error rates of 1 in 10^2 are experienced in both mediums. For successful CCS, an undetected error rate of around 1 in 10^{10} is required. Error control of CCS is therefore essential and constitutes the greatest source of complexity in the system.

In #6 (and #7) error detection is by redundant coding and error correction by re-transmission. There is therefore a requirement to maintain a re-transmission store at the transmit end, error-free message SUs being cleared from the store only on receipt of a positive acknowledgement signal. When an error is detected at the receive end, the corrupted SU is discarded. A negative acknowledgement bit, referring to the block containing the corrupted SU, is inserted in the appropriate place in the returned ASU.

The ASU (*fig. 7.19d*) contains:

Bits 1–3   011   ASU Header.
Bits 4–14   Eleven bits for eleven SUs set to 0, negative or 1, positive acknowledgement.
Bits 15–17   Sequence number of block which this ASU acknowledges.
Bits 18–20   Sequence number of the block containing the ASU.
Bits 21–28   Check bits.

The re-transmission procedure is initiated by recognition of a negative acknowledgement. The procedure varies according to the type of signal unit that has been corrupted:

LSU   A message LSU is re-transmitted.
MUM   When the corrupted SU is part of a message MUM, then the complete MUM is re-transmitted.
SYU   Corrupted idle SUs are not re-transmitted.
ASU   A received ASU is not acknowledged. If an ASU is corrupted it is assumed that the complete block is corrupt and all eleven bits of the returned ASU referring to that block will be negatively acknowledged. This means that, on occasion, unrequested re-transmission of good information will occur and an SU will be received error-free more than once.

Figure 7.20 illustrates this error correction process diagrammatically.

#6 signalling does not employ synchronisation of the "go" and "return" channels. For this reason and because of the nature of the error control mechanism, SUs and whole messages can be received out of order and more than once. The fixed position of the ASU every 12th signal unit means that it can break into a MUM which may be spread across more than one block. All these effects require the incorporation of complex "reasonableness" checks in the SPC processor using the CCS link. Thus, the processor is made complex and becomes involved in the signalling process, detracting from its available power to serve its proper functions.

CCITT Signalling System Number 7

The fact that #6 signalling used a 28-bit SU format which was inconsistent with the octal system adopted in digital PCM transmission is not actually a very serious drawback. #6 signalling is in use on 64 kbit/sec bearer channels using 4 "stuffing" bits which are of no significance to the message and are ignored. A signal channel capacity of 56 kbit/sec on a 64 kbit/sec bearer is therefore possible with #6 signalling.

The more serious drawback of #6 is that it places considerable additional responsibility upon the processor to sort out messages received twice or in unexpected sequence, etc. Co-incident with the specification of #6 in 1972

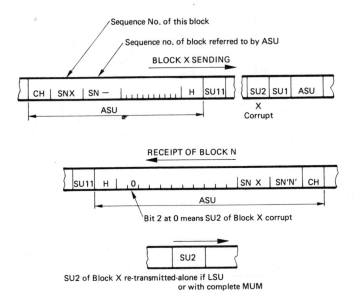

Fig. 7.20 CCITT No. 6 error correction

there was therefore international activity to define a CCS system more in keeping with the foreseen digital requirement, which would reduce or eliminate the need for processor activity and which would provide a signalling channel substantially independent of the nature of the messages being carried.

The System This last objective was conceived and formulated gradually as the definition of the future #7 system got under way. It is now central to the philosophy of #7, and to the international definition of open systems interconnection OSI, that the signalling system should be a bearer only. The signalling channel need know nothing of the nature of the message nor should the user processor need to know anything of the nature of the signalling system. In #7 therefore the system has a defined structure expressed in **levels** of increasing abstraction from the signalling link itself. Only level 4, the most remote from the signalling link, requires some recognition of the nature of the message user and #7 definitions include definitions of a level 4 Telephone Users Part, TUP, and Data Users Part, DUP. Other user part definitions at level 4 are envisaged.

This structural definition is illustrated in *fig. 7.21* and in more detail in *figs 7.22* and *7.23*, all of which are taken from the CCITT recommendations for #7 signalling.

The other fundamental concept adopted for #7 signalling was the concept of **variable message length**. Rather than perpetuate the need for multiple unit messages, inherent in the fixed word length structure of #6, it was decided to use a variable message length. This means that practically all messages are complete in themselves and not merely instalments of a total message. It means also that the error correction technique requires total change. #7 error correction is based on the use of forward and backward sequence numbers and forward and backward indicator bits. Use of these mechanisms

152 Introduction to Digital Communications Switching

Fig. 7.21 System No. 7 functional division concept

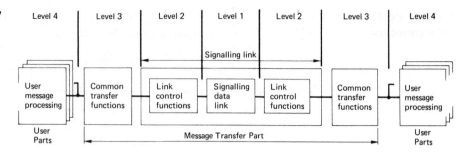

Fig. 7.22 System No. 7 conceptual structure

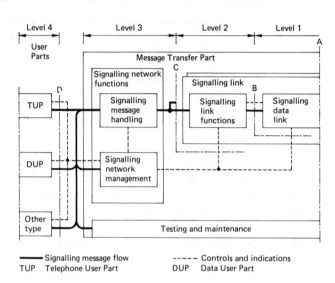

allows a protocol of cyclic re-transmission of all message data back to and including the SU found to be in error on receipt. The method is indeed flexible enough to allow a variety of error correction modes of operation. One such variety is *compelled error control* where no further messages are transmitted until the one just sent has been positively acknowledged. While this compelled mode grossly reduces the information capacity of the system it is of value in low-capacity applications (such as the concentrator to parent signalling of *fig. 5.27*).

Signal Unit Format Typical SU formats are illustrated in *fig. 7.24*. Because the SU is of variable length there is no need to define different messages in the same detail as in #6. Basically the SU consists of a portion concerned with the signalling function which is used by the message transfer part, MTP (levels 1, 2 and 3), and is not passed on to the user parts (level 4). The remainder of the message is passed direct to the user parts and is only considered by the MTP in determining the check digit pattern and the length information. Any message is made up of a number of octets. The number is variable between 0 and 62 for international use and between 0 and 272 for national

The Digital Network 153

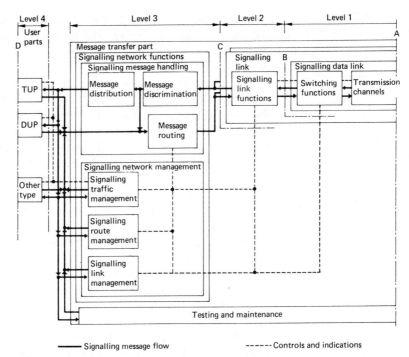

Fig. 7.23 System No. 7 structure in more detail

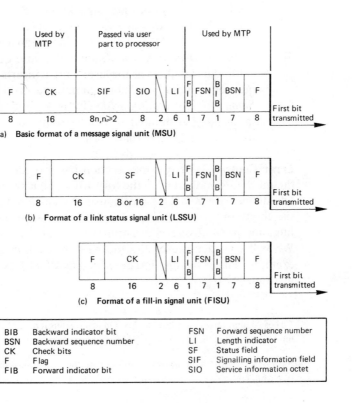

Fig. 7.24 System No. 7 basic signal unit formats

(a) Basic format of a message signal unit (MSU)

(b) Format of a link status signal unit (LSSU)

(c) Format of a fill-in signal unit (FISU)

BIB	Backward indicator bit		FSN	Forward sequence number
BSN	Backward sequence number		LI	Length indicator
CK	Check bits		SF	Status field
F	Flag		SIF	Signalling information field
FIB	Forward indicator bit		SIO	Service information octet

use. This limit excludes the number of octets concerned with the signalling portion of the message (*fig. 7.24*). The function of the signal unit portions defined in *fig. 7.24* is as follows:*

F   *Flag.* Defines the start of a message and the end of the preceding message. Has the code 01111110. To avoid imitation of the flag by the message contents, the transmit end inserts a 0 after every five consecutive 1s and the receive unit removes these protection zeros.†

BSN   *Backward sequence number.* Is assigned the same value as the FSN (see below) of the latest correctly received SU by the receiving end.

BIB   *Backward indicator bit.* Has its value inverted by the receiver end if the latest received SU is corrupted.

FSN   *Forward sequence number.* Is incremented by 1 each time a new message signal unit is being transmitted. Link Status Signal Units, LSSU, and Fill-in Signal Units, FISU, contain the same FSN as the last message SU transmitted.

FIB   *Forward indicator bit.* Is inverted by the transmit end when a re-transmission is commenced. Its value therefore becomes equal to that of the BIB which indicated re-transmission was required, starting from the message with FSN equal to the BSN plus one.

LI   *Length indicator.* Indicates the number of octets following LI and preceding the check bits. LI values can be interpreted as follows:
       LI = 0   Fill-in signal unit, FISU.
       LI = 1 or 2   Link status signal unit, LSSU.
       LI > 2 ≤ 62   Message signal unit, MSU.
       LI = 63   MSU with octets in the range 62 to 272.

SIO   *Service information octet.* Subdivided into quartets. The service indicator quartet defines the user part concerned. The sub-service field quartet defines international or national services.

SIF   *Signalling information field.* This is the body of the message expressed in an integral number of octets from 2 to 62 octets in length (international) or 2 to 272 (national).

Error Control   The previous section has provided a basic understanding of the mechanism inherent in the forward and backward sequence numbers and bits defined for #7. This mechanism, as indicated, allows some freedom in the choice of error control protocol. Cyclic re-transmission, compelled signalling, and preventive cyclic re-transmission are among the possible protocols. We will consider use of the mechanism using cyclic re-transmission as an example. This is the method described by CCITT as the basic error correcting method.

---

* These definitions are a trifle complex. On first reading it might be advisable to skip the definitions until after reading the section on Error Control.

† If HDB3 transmission protocol is in use then this zero insertion function is provided by HDB3 (see Chapter 3).

In **cyclic re-transmission**, the transmitted messages are stored. Storage capacity for 127 messages (of any length) is assumed. On receipt of a corrupted message, the receiver discards that message and all subsequent messages until re-transmission starts. The transmitter, on being told of an error, re-transmits the message that was corrupted and all subsequent messages.

Error Detection Prior to describing the error control process in more detail it will be as well to explain the error detection process. Errors in data sent on transmission links occur in bursts. Theory and practice have shown that *polynomial check codes* are the most proficient for detecting such errors. Unfortunately, although it is possible to detect all combinations of an odd number of errors in a message, no polynomial of practical dimensions is capable of detecting all combinations of an even number of errors in a message. However, it is possible to select a polynomial which would fail on combinations which are extremely unlikely to occur, particularly for the lengths of message which may be transmitted. In #7 with a maximum, for international operation, of 62 octets which have to be protected and using 64 kbit/sec as the transmission means, the chosen polynomial is

$$x^{16} + x^{12} + x^5 + 1$$

Figure 7.25 illustrates a practical arrangement for error detection using the recommended polynomial. When a binary stream of data is passed through the feedback shift register, the practical effect is to divide the binary value of the given data by the chosen polynomial and the final value left in the shift register is the remainder of this division. This remainder is transmitted immediately following the original data and the entire bit stream is treated by a similar shift register at the distant terminal.

If the initial state of the receiving shift register is all-zero and there are no transmission errors, the contents of the shift register will pass through a number of combinations dependent upon the received data and will finally return to its original all-zero state. However, it is possible that the final all-zero

Fig. 7.25 System No. 7 type of error detection implementation

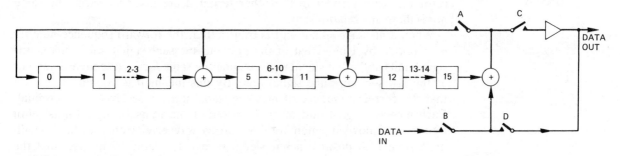

Shift registers in coders are "all 1" at start

Transmitting terminal
When information bits are being transmitted — Switches A, B and D closed, C open
When check bits are being transmitted — Switches A, B and D open, C closed

Receiving terminal
When information bits are being received — Switches A, B and D closed, C open
When check bits are being received — Switches A and B closed, C and D open

condition could be due to causes other than the absence of transmission errors, for example the shift register itself may be faulty. In #7 it is recommended that, to guard against such malfunctions, the initial condition of the transmitting and receiving shift registers is the all-one state and also the remainder in the transmitting shift register is sent as the ones complement. (This means that all bits are inverted before transmission, that is 1 changes to 0 and vice versa.) When there are no transmission errors, the final state of the shift register is a definite pattern, irrespective of the data; any other pattern denotes a transmission error. Signal units which pass the error detection test go forward to the error correction phase and the others are discarded.

Error Correction—Basic Method The basic error correction method makes use of the forward sequence number, the forward indicator bit, the backward sequence number, and the backward indicator bit. Inspection of these results in a positive acknowledgement for correct receipt of information and a negative acknowledgement for a corrupt transfer. Also, the check of the forward sequence number ensures correct signal unit sequence. An important point to remember is that the forward sequence number and the backward sequence number in a particular signal unit are not related; the forward sequence number and forward indicator bit in signal units in one direction are associated with the backward sequence number and backward indicator bit in signal units in the reverse direction for carrying out error correction in the first direction. Also error correction at the two terminals of a signalling link take place independently. This is made clear in the explanatory diagrams of *fig. 7.26*.

Initially, the forward and backward indicator bits of signal units in the two directions are the same, say 1. At a particular time after a sequence of correct signal units, the backward indicator bit received at a terminal will be the same as the forward indicator bit which has been sent in the last transmitted signal unit. Because of previous events, the forward and backward indicator bits with respect to the other terminal are not necessarily the same nor is there any relationship.

When a message signal unit is transmitted, the forward sequence number is increased by 1 over that of the previous message signal unit. Link status signal units and fill-in signal units assume the same forward sequence number as the last sent message signal unit. By this means it is possible to ensure that the correct sequence of message signal units is received at a terminal. Each message signal unit must be stored at the transmitting terminal until a positive acknowledgement for this message is received from the other signalling terminal. A positive acknowledgement is indicated by the fact that the signal unit from the receiving terminal contains a backward indicator bit which is the same as that in the previous message from the receiving terminal, and the message to which this positive acknowledgement refers is denoted by the backward sequence number. The information returned from the receiving terminal to the original transmitting terminal may be contained in any of the three types of signal units. When a transmitting terminal receives a positive

The Digital Network 157

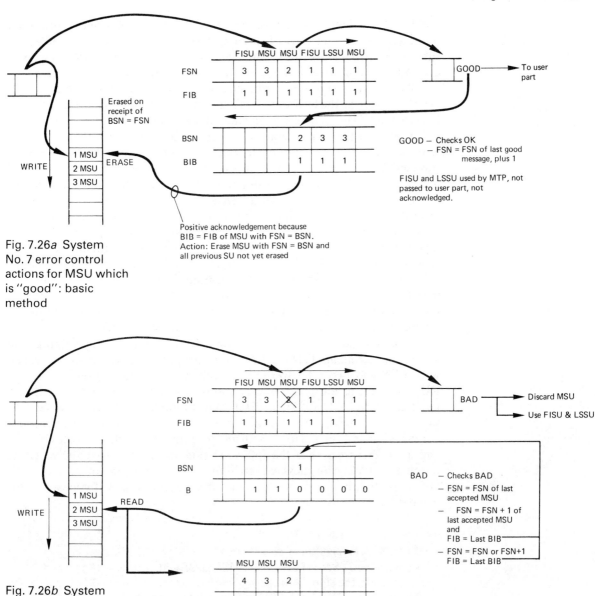

Fig. 7.26a System No. 7 error control actions for MSU which is "good": basic method

Fig. 7.26b System No. 7 error control actions for MSU which is "bad": basic method

acknowledgement for a message signal unit, this also represents an acknowledgement of all previous message signal units which may not have been separately acknowledged. No positive acknowledgements are required for link status signal units and fill-in signal units, which can be recognised by their length indicators.

When a signal unit is received in a corrupt form, this signal unit is discarded at the error detection stage. If it happened to be a message signal unit, the subsequent signal units would contain unexpected forward sequence numbers even though they may have passed the error detection test. The receiving terminal now has to respond with a negative acknowledgement which is indicated by inverting the backward indicator bit from that of its last sent backward indicator bit. The backward sequence number in the corresponding signal unit is that of the last accepted message signal unit. When this signal unit is received at the transmitting terminal, the backward indicator bit being reversed from its previous state is also in the inverse state to the forward indicator bit which the transmitting terminal has last sent. Thus the transmitting terminal recognises that re-transmission is necessary, and this must start from the message signal unit whose forward sequence number is one more than the backward sequence number in the negative acknowledgement signal unit from the receiving terminal. All message signal units with higher forward sequence numbers which are still waiting in the store until positively acknowledged are also re-transmitted. The re-transmission process is indicated by inverting the forward indicator bit so that once again it becomes equal to the backward indicator bit in signal units returning from the receiving terminal, i.e. the backward indicator bit which initiated the re-transmit action.

This description has referred continually to a transmitting terminal and a receiving terminal. This has been in respect to MSU sent in one direction only. FSN and FIB refer to the MSU of which they form a part. BSN and BIB refer not at all to the message in which they are carried (which may be of any type MSU, LSSU or FISU) but to the last received message. The same mechanism is in operation in respect of messages in the reverse direction.

Figure 7.26 is designed to assist in understanding this rather complex process and as further assistance the following summarises events at the receiving terminal, not only for message signal units but also for the other two types. Differentiation between messages of different type is made by means of the length indicator.

1   *LI indicates reception of an MSU*.
    a) Received FSN = FSN of last MSU accepted.
       Discard the signal unit.
    b) Received FSN = 1 + FSN of last MSU accepted.
       i) Received FIB same as last sent BIB.
          Accept signal unit and pass to level 3.
          Send positive acknowledgement.
       ii) Received FIB different from last sent BIB.
          Discard the signal unit.
    c) Received FSN is not as in *a*) or *b*) above.
       Discard the signal unit.
       If received FIB same as last sent BIB, send negative acknowledgement.
2   *LI indicates reception of an LSSU*.
   Process MTP and status field information.

3 *LI indicates reception of an FISU.*
   a) Received FSN = FSN of last MSU accepted.
      Process MTP.
   b) Received FSN different from FSN of last MSU accepted.
      Process MTP.
      If received FIB same as last sent BIB, send negative acknowledgement.

For all of the above, the backward sequence number and the backward indicator bit must be monitored to deal with error correction of signal units sent in the reverse direction.

At the transmitting terminal, reception of a positive acknowledgement for a particular message signal unit denoted by the received backward sequence number is taken as an acknowledgement of all previously sent message signal units even though the corresponding backward sequence numbers have not been received. This means that all such signal units can be removed from store. When a negative acknowledgement is received, the message signal unit denoted by the backward sequence number plus 1, together with all subsequent message signal units which have been sent, are re-transmitted using their original forward sequence numbers. New message signal units are not sent until after the re-transmission of the last message signal unit available for re-transmission.

Error Correction—Preventive Cyclic Re-transmission Method The CCITT defines an alternative error correction method in addition to the basic method. This is known as preventive cyclic re-transmission.

Error correction by preventive cyclic re-transmission is based upon positive acknowledgements only and uses forward and backward sequence numbers but not forward and backward indicator bits. The latter are pre-set to a chosen condition (the CCITT recommends 1) but they play no part in the operations to be described.

As in the basic error correction method, when a message signal unit is transmitted, the forward sequence number is increased by 1 over that of the previous message signal unit, and all message signal units are retained until a positive acknowledgement is sent back from the receiving terminal. Link status signal units and fill-in signal units assume the same forward sequence number as the last transmitted message signal unit.

Preventive cyclic re-transmission differs from the basic error correction method in that, when there are no new message signal units or link status signal units awaiting transmission, re-transmission is commenced, starting with the message with the lowest forward sequence number of all message signals still being stored whilst waiting for acknowledgement. The original forward sequence numbers are maintained. If during the re-transmission phase a new signal unit becomes available, the re-transmission cycle is interrupted and the new unit is transmitted, after which the cycle is continued. The re-transmission cycle continues until a positive acknowledgement is received when, in consequence, those signal units having a forward sequence number equal to or less than the backward sequence number in the positive acknowledgement message are removed from store.

Positive acknowledgement is indicated to the transmit end by receipt of the BSN equal to the FSN related to the message being acknowledged.

When all transmitted signal units have been given a positive acknowledgement and no signal units have to be transmitted, fill-in signal units are sent.

Using the preventive cyclic re-transmission procedure it is possible for the holding store to be filled and, if not prevented, information lost. To prevent this there has to be a limit to the number of message signal units or the number of message signal unit octets available for re-transmission. When either of these limits is reached, a procedure known as *forced re-transmission* is adopted. In this forced re-transmission process, no new message signal units or fill-in signal units are sent and all the message signal units available for re-transmission are re-transmitted with priority, in the order in which they were originally transmitted. This operation continues until, at the end of a cycle of forced re-transmission, the store contents are less than the two permitted limits. Then the normal preventive cyclic re-transmission operation is resumed.

Preventive cyclic re-transmission is recommended for applications where the propagation delay of the link is long (satellite circuits, for example). In such circumstances, the re-transmission on negative acknowledgement of the basic method introduces unacceptably long delays.

With CCITT Signalling System Number 7 we have, for the first time in the history of communications switching, a method of signalling quite independent of the user and of the message content. This powerful new tool means that, in theory, all message communication between processors, and between machines belonging to end users with processor capability, can use the same message channels. This, as we have noted, is not to be the case in the UK at least but even the alternative CCS systems devised to suit particular interests and groupings are being formulated using the same principles as #7.

Open Systems Interconnection— the ISO Reference Model

It is often the case that a technological breakthrough, once achieved, impresses the beholder by its simplicity. "Why did nobody think of that before" can be a typical reaction. The concept of PCM itself and certainly the concept of levels of abstraction from the communications link inherent in #7 are two such advances, obvious with hindsight. The #7 signalling concept of levels may have been instrumental in formulating a more fundamental and far-ranging concept applicable to all machine communications. This is the ISO Reference Model for Open Systems Interconnection proposed by the International Standards Organisation in 1979 and adopted by the CCITT thereafter.

The object of the ISO model, similar to but more fundamental than #7 signalling, is to allow diverse processors to communicate via diverse networks and communications links without the need for the processor to recognise the special peculiarities of the link or vice versa.

To achieve this aim the model defines seven layers. Each layer communicates with layers of equal order via communication protocols special to that layer. The actual communication passes, via defined interfaces between the

Fig. 7.27 Principle of layered model

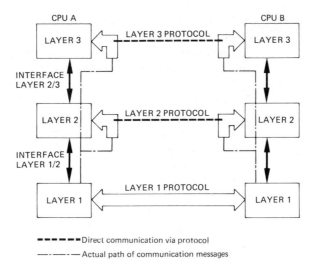

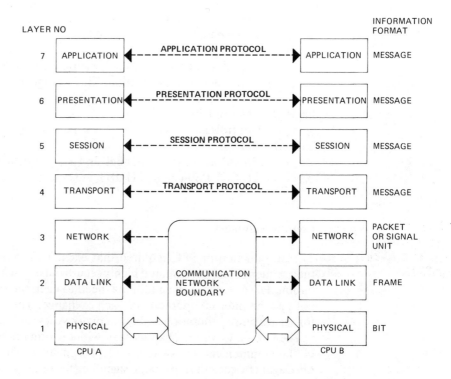

Fig. 7.28 ISO Open Systems Interconnection reference model

layers, down to the lowest layer, which is concerned with the link itself, and thence via the link and upwards to the target layer (*fig. 7.27*).

The full seven-layer model is shown in *fig. 7.28* from which it can be recognised that #7 signalling approximates to layers 1 to 3 of the model. The functions of the layers are as follows:

Layer 7 APPLICATION
Directly serving the end user.
Example. In a distributed database this layer could instigate the interrogation of several host computers.

Layer 6 PRESENTATION
Provides services to the application layer to enable it to interpret the data. For example,
encryption
code conversion
compression
display formatting.

Layer 5 SESSION
Session administration, e.g. authenticating the dialogue between two parties (processors).
Session dialogue: collating and assembling messages associated with a particular dialogue.

Layer 4 TRANSPORT
Provides transparent transfer of data between session entities.
Relieves the session layer from any concern with the method by which data transfer is achieved.

Layer 3 NETWORK
Provides functional and procedural means to exchange information over a network connection. This will include routing and switching information.

Layer 2 DATA LINK
Controls and ensures error-free data exchange.

Layer 1 PHYSICAL
Provides the data channel and its protocol, e.g. CCITT V24 or 21, CEPT 30-channel HDB3, etc.

Chapter Summary

The introduction of SPC required that exchanges should operate to a fixed time sequence and necessitated highly secure pulse generators and pulse distribution. PCM imposed the additional need to keep the network synchronised, not just the internal operation of each exchange. The resulting requirements for secure pulse generation and for network synchronisation were described.

The remainder of the chapter dealt with the other network related function of the communications switch, that of signalling. The background was set once again by considering analog signalling techniques so that the new requirement of digital transmission and switching can be seen in context.

Digital channel associated signalling was considered only in relation to the CEPT 30-channel PCM system.

An extended treatment was given to the concepts of common channel signalling followed by the details of the two such systems so far defined and in use: CCITT Signalling System No. 6, a fixed word length, fixed frame system; and CCITT No. 7, a variable word length system which renders the communicating processors largely unconscious of the signalling mechanism and medium.

#7 signalling first introduced the concept of separating communicating processors from the communications medium by layers of abstraction and the final part of the chapter dealt with the ISO OSI model which generalises this concept for any communication between any processor type entity.

References

[7.1] *Post Office Electrical Engineers Journal*, Vol. 72, p. 132, July 1978.
[7.2] *Post Office Electrical Engineers Journal*, Vol. 70, p. 21, April 1977.
[7.3] *Post Office Electrical Engineers Journal*, Vol. 73, p. 88, July 1980.
[7.4] *British Telecommunications Engineering*, Vol. 3, p. 99, July 1984.
[7.5] *Digital Telephony: An Introduction*, an Ericsson book (Dkt127BUe 102 721 Ue January 1977).
(This excellent work is now out of print but very helpful and still relevant.)
[7.6] *Topic 7: Digital Telephony*, Siemens Aktiengesellschaft (Booklet No. N100/3117.101).
(Failing to obtain reference 5, this is an ideal substitute.)
[7.7] S Welch, *Signalling in Telecommunications Networks* (Peter Peregrinus 1979).
[7.8] *CCITT Yellow Book*, Vol. V1, Fascicle V1.4, 1980.
(R1 and R2 signalling.)
[7.9] *CCITT Yellow Book*, Vol. V1, Fascicle V1.3, 1980.
(Signalling System Number 6.)
[7.10] *CCITT Yellow Book*, Vol. V1, Fascicle V1.6, 1980.
(Signalling System Number 7.)
[7.11] P Bylanski and D C W Ingram, *Digital Transmission Systems* (Peter Peregrinus 1976).
Note: CCITT Red Books 1984 replaced the Yellow Books when published during 1985.

Exercises 7

7.1 List the timing functions required in an automatic telephone exchange regardless of its technology.

7.2 Classify the timing functions required in an SPC digital exchange into
   a) Those independent of GMT and the network
   b) Those dependent on the network
   c) Those dependent on GMT.

7.3 Assuming that the digital exchange is responsible for HDB3 transmission coding, what is the highest frequency required from the clock? You may ignore control timing which may well be based on a much higher frequency.

7.4 Assuming the fundamental frequency is as determined in Exercise 7.3, list the divisions required to produce lower frequencies for other CEPT 30-channel PCM functions.

7.5 Define in your own words what is meant by
   a) Plesiochronous operation
   b) Synchronous operation
   of a network timing system.

7.6 What will be the effect of slip on
   a) Voice communication
   b) Facsimile transmission
   c) A viewdata communication.

7.7 Define the concept of error-free one-second intervals, EFS, in data communications. What will be the number of slips per day in a 1544 kbit/sec system using a frequency accuracy of 10^{-10}?

7.8 Why have register signals remained for so long as loop disconnect whereas line signalling adopted AC techniques much earlier?

7.9 What considerations govern the choice between inband and outband AC line signalling?

7.10 In an imaginary CEPT 30-channel channel-associated signalling system, a signal is accepted if it persists for 10 frames and is not longer than 30 frames. Ignoring transmission delays what is the minimum delay between start of sending and signal recognition?

7.11 What new functions have to be introduced to the exchange system because of the use of CCS?

7.12 If a CCS link is used to serve 1500 traffic circuits, how many subscribers would be affected by its loss in the busy hour? Subscribers originate 0.04 E per line and traffic circuits are loaded to 0.85 E.

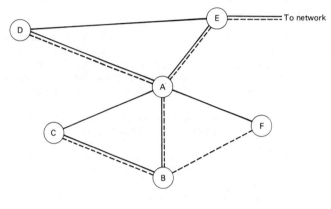

Fig. 7.29 CCS network

7.13 *Figure 7.29* depicts a network using CCS. Categorise the links listed below as Associated, Quasi-associated, etc.:

A–F	C–A
A–B	B–D
D–network	D–C

7.14 Write down the essay plan for an essay on CCS security back-up architectures with reference to the UK and USA approaches.

7.15 List the basic differences between #6 and #7 signalling.

7.16 Try to adduce a reason for each of the differences listed in your answer to Exercise 7.15.

7.17 What is the error control action of #6 signalling if an ASU is received corrupted? Comment upon the problems associated with this procedure.

7.18 Assuming that messages are LSU only, what is the message information capacity of a #6 signalling link sent over a 64 kbit/sec bearer channel?

7.19 In #7 signalling what effect does the insertion of zeros to prevent flag imitation have on the transmission channel capacity?

7.20 Ignoring the effect noted in Exercise 7.19 calculate the message information capacity of a #7 link assuming all messages consist of 2 octets and only MSU are sent. (If you are interested in comparing your answer to that of Exercise 7.18, note that 2 octets represent 4 times the message capacity of a #6 LSU.)

8 Digital Frontiers

Digital communication is only suitable for communications between machines; human beings are limited to analog methods. Therefore, somewhere in the digital network there will be interfaces, either permanent for telephony-only subscribers, or temporary for the telephonic human intervention into machine-to-machine communications.

Fig. 8.1

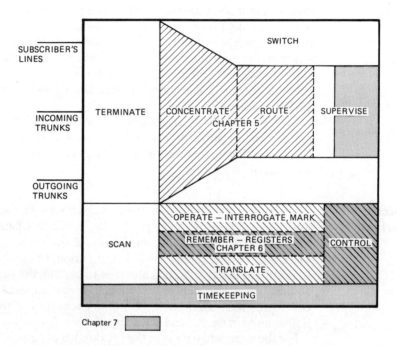

The familiar introductory diagram is presented in another form in *fig. 8.1*, this time showing the topics covered by the preceding chapters. We have still to cover the frontier areas of the exchange where this interface is probably located, the line and trunk terminations and the scanning function. Inherent in the discussion of lines and trunks is their testing for continued integrity.

In considering the subscriber's line interface it is appropriate to include what little can be said about the future of integrated voice and data over the subscribers' network to the digital subscriber's terminal. Another kind of frontier area is the physical location of the exchange equipment within the building. It is the saving in building costs by increasing miniaturisation that has provided much of the impetus to the development of electronic solid state exchange systems.

Line and Trunk Scanning

Analog exchange systems required a separate scanning function. The Strowger line circuit used reversed scanning; each individual calling condition caused a uniselector to scan for a free first selector. Marker control systems used some form of periodic scan of calling leads from each line and incoming trunk. Later, SPC exchanges employed a similar technique.

In all these cases the periodicity of the scan of subscribers' lines is determined by the tolerable duration of the silence prior to receiving dial tone. Relatively slow scans of 100 or 200 msec are adequate. For incoming trunks a similar period is acceptable in register controlled systems when the transmission of digits awaits a proceed-to-send signal. However, a shorter period will improve post-dialling delay. In some systems, however, notably the UK step-by-step system, digits are transmitted without a proceed-to-send signal after the end of the inter-digit pause. In these circumstances the incoming trunk scan must be rapid enough to detect a calling signal in less than 50 msec, the only margin left from the inter-digit pause after allowance for connection times. This problem of fast scan of incoming trunks will remain in the UK mixed analog/digital network for some considerable time since more modern systems, notably TXE 2, employ similar arrangements.

In digital systems the scanning function becomes inherent in the multiplexing process and requires little further discussion. It is still necessary to ensure that the calling indications result in the appropriate connection actions within the prescribed periods.

The Subscriber's Line Interface

An introductory discussion of the subscriber's line interface was provided in Chapter 5 (see *fig. 5.20*, p. 89). We noted in Chapter 5 how a very high proportion of the cost and complexity of the exchange has been moved by successive developments towards the periphery and beyond. This tendency is illustrated in *fig. 5.20*. It is also true to say that the periphery, the telephone and the local line network particularly, has imposed constraints upon the development of solid state electronic exchanges. This part of the problem is illustrated in *fig. 8.2* and is worthy of some discussion.

For the complete network, the breakdown of costs is very approximately:

	Analog %	Digital %
Switching	45	45
Signalling	15	10
Transmission	10	5
Terminals	15	30
Local line plant	15	10
	100	100

The second column is shown on the assumption that the digital interface is in the subscriber's premises or the 'telephone' instrument itself (*fig. 5.20*). Of this cost breakdown, the only item which is paid for by the subscriber

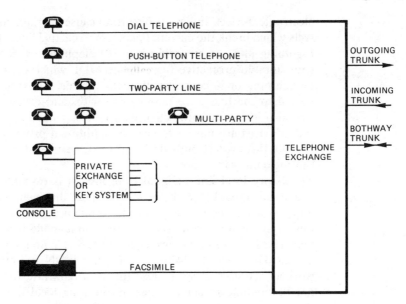

Fig. 8.2 Telephone exchange subscribers

immediately on installation is the terminal; all the other items represent investment well in advance of revenue returns. It is, therefore, in the interests of the operating company to move towards the digital integrated network. The 30% of the analog investment contained in the terminal and local line plant is far too big a proportion for it to be practical to make this plant redundant by the introduction of a new system. The new digital system must, therefore, interwork with the existing local network (*fig. 8.2*).

Interface with the Existing Local Network

Several features of the existing telephone plant provide problems to the digital system designer. These features are as follows in decreasing order of importance:
 The telephone bell.
 The DC feed requirement to the transmitter.
 The incidence of lightning and power cross-connection on the local line plant.
 The 2-wire local line.

The telephone bell The characteristics of the subscriber calling device were determined by the nature of the magneto caller in the earlier manual systems. The bell or modern tone sounder, therefore, responds to a ringing current of 15–25 Hz at a voltage of about 75 V rms with a power requirement in excess of 1 watt. These are characteristics which are alien to solid state electronic circuitry.

The dc transmitter feed The carbon microphone requires some 25 mA at the nominal 50 V exchange battery voltage. Again such values are alien to electronic devices and are sufficient to cause sparking at miniature relay contacts which break the current flowing in a reactive line.

Lightning and power accidents All telephone exchanges have been fitted with suitable protective fuses at the MDF which disconnect the line struck by lightning or in contact with mains power. Electromechanical relays connected to the line were designed to withstand excess surges in the interim period prior to the protective devices disconnecting the line. Modern solid state devices are not so robust and additional protection must be included. Usually this protection is designed to be destroyed by the surge but without causing danger from fire.

The 2-wire local line Digital transmission is essentially 4-wire and to this extent incompatible with the 2-wire local line plant. Export of the digital interface from the exchange to the terminal therefore requires a solution to this problem. One solution is to use **burst mode transmission** on the local line. This is illustrated in principle in *fig. 8.4*. With the CCITT recommending a total data rate of 144 kbit/sec for the ISDN subscriber then burst mode working requires at least twice this as a data rate for the local line. It will not be possible to achieve this on the long local lines common in existing networks without incorporating some form of repeater. The maximum range achievable on local line plant without repeaters will be less than 5 km.

A more satisfactory solution may be to employ a **digital hybrid**. In just the same way as the analog 2-wire telephone service employs 4-wire to 2-wire termination in the telephone set and at the entry to the 4-wire trunk network, so too a similar function can be provided at the digital instrument and at the exchange digital line interface.

Figure 8.3 illustrates the new devices which will share the new digital network, adding to the problems of compatibility with existing local line and subscribers' apparatus. These are the devices whose connection has been made possible by the digital interface moving from the exchange to the subscriber's premises, creating the Integrated Services Digital Network, ISDN.

It is the problem of compatibility with existing devices which has proved the most intractable in introducing electronic solid state telephone exchange systems. Reasonably satisfactory, economical solutions have required the development of customised large-scale integrated circuits. The line interface circuits in use today are, in general, interim solutions and development continues. It is still difficult to meet all the CCITT requirements for a subscriber's interface and in most cases unsatisfactory compromises have to be accepted.

Digital Frontiers 169

Fig. 8.3
Communications
switch subscribers

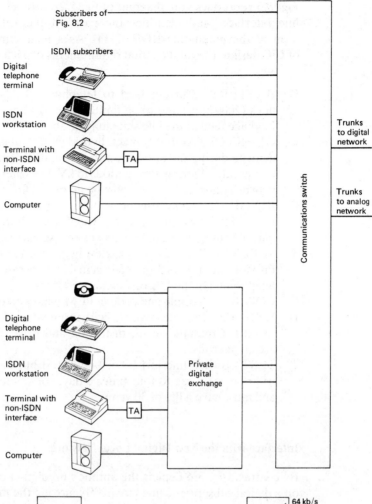

Fig. 8.4 Principle of
burst mode working

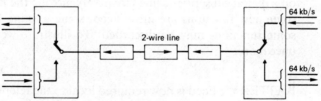

Note: Switch must operate at a high-frequency sufficient
to pack continuous 64 kbit/sec data onto burst "packets"

Fig. 8.5 reproduces the diagram of *fig. 5.21*, a block diagram of the AXE 10 line interface, and identifies the parts of the LIC associated with each term in the mnemonic BORSCHT. As a final summary of the difficulties of LIC design, the fundamental requirements of each item are listed:

B BATTERY   Current feed to telephone. To be about 25 mA on long lines (2000 ohm) and to be limited on short lines to little more than this. Leakage feed in the idle state may be about 1 mA.

O OVERVOLTAGE   Protect hybrid and remaining LIC circuits from damage during the period (up to $\frac{1}{2}$ second) that the MDF protectors take to operate. During this period 15 kV low energy or 415 V rms power may be connected. There must be no risk of fire.

R RINGING   Apply ringing voltage and cadence to line, typically 75 V rms 15 Hz interrupted ringing. Ringing can hurt the human listener and must be cut off immediately an answer (seizure) is detected.

S SUPERVISION   Detect seizure by loop *even during the ringing signal*. Provide line reversal or other signals to operate charge recorders at the telephone. (In the UK this signal is 50 Hz tone.)

C CODING   Sample and code in PCM binary octet, 7 bits plus sign bit.

H HYBRID   Provide 4-wire/2-wire termination. Loss across hybrid limited to 1 dB. Crosstalk, echo, distortion and other transmission parameters are all onerous.

T TEST   Switch line to LIC or line to test bus or LIC to test bus. Relays will operate three to four times a day, sometimes switching line current, and must have a life of 30 years.

Interface with the New Digital Local Network

By contrast, *fig. 8.6* depicts the author's prediction of the functional division on a digital subscriber's line circuit. By moving the interface to the telephone some new functions are introduced, some existing functions simplified, and some functions may be discarded. To illustrate we will go through the list once more:

BATTERY   Feed is now required local to the telephone. This could be from a local power source but is more likely to be a DC line feed extracted at the telephone. *Figure 8.5* does not show how this is done.

OVERVOLTAGE   Protection is now required at both ends of the line. Previously the instrument was protected only by a simple fuse but there is now sufficient investment in the instrument to warrant protection from destruction.

RINGING   The calling signal is produced locally from the decoded call signal. The compatibility problem has been removed.

SUPERVISION   This is by coded signals and requires little that is additional to the PCM facilities.

CODING   Coding takes place in the telephone.

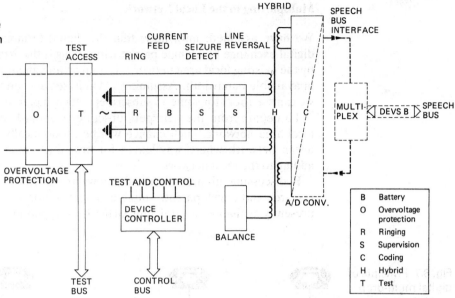

Fig. 8.5 Functional diagram of digital line interface: AXE system

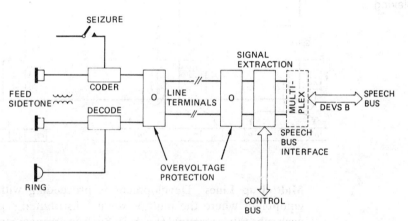

Fig. 8.6 Functional diagram of digital subscriber line interfaces

HYBRID/SIDETONE No longer required but with the removal of the hybrid previously in the telephone some means has to be added to provide the sidetone which was caused by speech and noise leaking from go-to-return paths of the hybrid. Without sidetone the subscriber hears silence on attempting a call and therefore doubts if the exchange is "there"; also the subjective quality of the completed call is unsatisfactory.

TEST The test function is greatly simplified. The line is now passing PCM streams even when idle and these can be used for test purposes. There is no longer a need to measure transmission parameters of the line. Whatever testing may be required can probably be effected over the PCM stream.

LINE TERMINAL A new function has appeared in the requirement to adopt a 4-wire protocol for a 2-wire line. Such a protocol as burst mode has been suggested. The line terminal will at least include this function and may also include a feature to "multi-drop" (see below).

Multiplexing in the Local Network

We have assumed, until now, that the digital terminal is connected to the digital exchange by a single pair of wires. This is the arrangement encouraged by the existing local area network. It is an arrangement that is grossly wasteful in available bandwidth. Assuming 144 kbit/sec allocation per subscriber then we can see room for some 15 subscribers on each pair of wires.

To depart from the pair per subscriber standard requires change to the local area network. Such change will be economically justifiable first in the urban network where way-leaves for new duct routes escalate the cost of adding to the local network.

In discussing this area of the subject we are moving into the field of ongoing research and development. Suitable system solutions are not always or often presently available. Several approaches to solutions are perceivable.

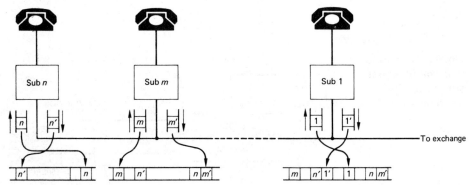

Fig. 8.7 Principle of digital multi-drop multiplexing

Multi-drop Lines Development is proceeding with 30-channel multiplex equipment where the multiplexer is "distributed", part being contained in each subscriber terminal (*fig. 8.7*). Such an arrangement necessitates the provision of two pairs of wires from the distributed multiplexor to each terminal and requires design solutions to the security problem of the subscriber being dependent for service on equipment located in another subscriber's terminal.

Common Service Distribution The provision of cable television necessitates a new local distribution of high-capacity coaxial or fibre optic bearers to each (CTV) subscriber's premises. Assuming widespread acceptance of this technique then the communications' services could "piggyback" on the cable TV bearer. Other services such as gas, water and electricity could then use the new communications service to read meters and even, perhaps, to bill and collect using electronic funds transfer.

This solution is applicable in the first instance to suburban areas, particularly areas with a large amount of new housing where the new bearers are provided as part of the building process. It does not assist in the urban areas first requiring local area multiplexing.

Low Bit Rates Voice communication involves about 100 significant bits of information per second so that the provision of a 64 kbit/sec bearer represents a degree of overkill. The reasons for this wide bandwidth provision are associated with the problems of digital transmission in a mixed analog/digital network. When the network has become all digital then the need for 64 kbit/sec will no longer exist. It is predictable, therefore, that developments will emerge which use lower bit rate coding techniques for some or all of the connection. This can be accomplished by sending a class mark with each connection, the class mark being used to allocate the appropriate channel capacity to the connection. It is evident that a lower bit rate does not solve the local network problems of sharing pairs between subscribers but it may alleviate the problems involved in the solution.

Local Switching The solution to the problem in the urban areas may lie, in part, in the proliferation of private exchanges. Although digital PABXs are entering service in large numbers, there are, at the time of writing, no digital PABXs using a 64 kbit/sec exchange line interface to the public network. In the first instance a 2.048 Mbit/sec interface will be made available in the UK. This will use DASS CCS. Thirty exchange lines represents rather too large a capacity for the majority of small PABXs. The small PABX does, however, offer a suitable home for the siting of multi-drop equipment which could, in this application, be made more secure by extending the line into a ring and using a ring-type protocol similar to those coming into service in local area network applications for computer processor communications.

The extension of the PABX into the public domain is the remote concentrator and small supervised exchanges, both of which were discussed in Chapter 5.

The Trunk Network Interface

The trunk interface with the digital network has been dealt with adequately already. We have neglected the interface with the existing analog network.

Where the digital exchange requires trunks to an existing analog exchange, these may be provided either as analog trunks, when the coder will be located at the digital trunk interface, or as digital trunks, when the coder will be located with the analog exchange (*fig. 8.8*). The latter alternative will be more economic and make the eventual replacement of the analog exchange more simple but may require non-existent space and facilities at the analog exchange.

In either case, signalling conversion will be a function of the digital exchange and will introduce a number of problems. Should the digital network be extensive and use CCS then this analog route will represent a problem. Even if signal conversion to CCS is used for onward connections, the signalling data available from the analog route will be limited. Alternatively, the analog route may be made to dictate onward channel associated signalling over the digital network. Again, for this alternative, the digital network must be made to modify its behaviour to suit.

Figure 8.8 illustrates a typical development scenario for a network moving from analog to digital. Provided that the network develops in this gradual manner then the signalling problems discussed may not arise. In the early phases of digital development, channel associated signalling will be the norm and CCS when it is introduced will be confined to the digital portion of the network. We will, however, have imposed additional signalling requirements upon the digital exchange which will be unnecessary when the network becomes all digital.

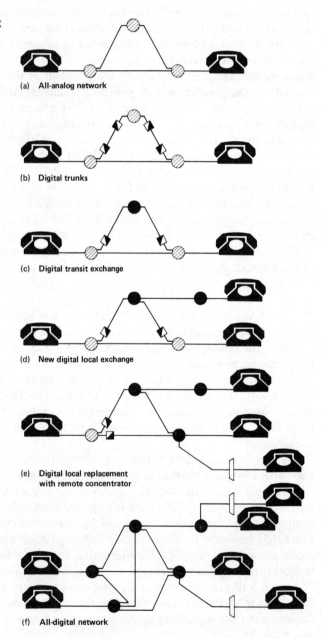

Fig. 8.8 Development of a digital network

(a) All-analog network

(b) Digital trunks

(c) Digital transit exchange

(d) New digital local exchange

(e) Digital local replacement with remote concentrator

(f) All-digital network

Line and Trunk Testing

A feature that has been treated in vastly different ways by different administrations and operating companies has been the testing of lines and trunks. At one end of the scale, the UK has, until very recently, relied on the subscriber to notify line faults and used a sophisticated manual test desk to establish the nature of the fault. For trunk testing, the UK has relied upon routine testing of the trunk terminations, a permanent backward signal to indicate that the trunk is connected, and manual tests. By contrast, the German administration has always insisted upon routine, automatic testing of all functions (transmission quality and signalling) of all lines and trunks.

More recently, administrations have been encouraged both by the availability of "add-on" routine test equipments and by the increasing costs of labour to introduce automatic routine testing. Every exchange system, therefore, offers as standard the test access relay arrangement described in Chapter 5 (*fig. 5.25*) and most systems include routine test features. Similar arrangements are included for analog trunks.

Testing of digital trunks is more debatable. The continuing connection of the trunk is indicated by the continuous reception of the PCM bit stream. Test patterns incorporated in channel 0 further check the integrity of the common portions of the coder continuously. There is a remaining section of the trunk which is not checked for integrity, the circuitry associated with the individual channel. In most circumstances, routine testing of this remainder is not thought necessary. Occasional channel testing can be accomplished by using the loop-back features standard on PCM repeaters and terminals.

Integrated Networks

To understand present-day intentions for the digital network it is necessary to have an understanding of what is meant by the abbreviations being used to describe them.

Digital switching allied to digital transmission results in an *integrated digital network*, IDN, where there is no longer any distinction between "outside plant" (transmission and bearer channel or circuit) and "inside plant" (switching). The IDN will exist when all exchanges are digital and all links between exchanges use digital coding.

Once a digital interface is moved to the subscriber's premises then the conditions exist for the subscriber's telephony, data processing, telemetry, viewdata and other needs to be met using the one digital bearer. To provide convenient service, the data-type transactions should not be interrupted by speech telephony or telemetry requests. The bearer circuit per subscriber is, therefore, intended to be more than one channel shared between services. CCITT recommends

$2B + D$

with B_1 64 kbit/sec voice; B_2 64 kbit/sec data; D 16 kbit/sec data, telemetry or signalling.

Where this multi-channel access to the subscriber is provided then we describe the resulting network as the *integrated services digital network*, ISDN.

In the UK, an ISDN tral was planned on the basis of a less-generous provision of

2B + D channels
with B_1 64 kbit/sec; B_2 8 kbit/sec and D 8 kbit/sec

with digital access to PABXs at 2.048 Mbit/sec using DASS signalling. This trial, after many postponements, commenced in mid-1985. The ISDN trial in the UK has been christened *integrated digital access*, IDA.

Summarising:

IDN **Integrated Digital Network**
A combination of digital nodes and digital links that uses integrated digital transmission and switching to provide digital connections between two or more points to facilitate telecommunications and, possibly, other functions.

ISDN **Integrated Services Digital network**
An integrated services network that provides digital connections between user network interfaces in order to provide or support a range of different telecommunications services.

IDA **Integrated Digital Access**
A network providing integrated access supporting a number of diverse services to user network interfaces from the integrated digital network. Used specifically to describe UK field trials.

At present the ISDN is of the future. Apart from the UK field trial mentioned above, Ericsson have conducted one or two partial trials in Sweden and are at present engaged in a more comprehensive trial in Venice. The French PTT also have trials and customer acceptance tests in progress on some elements of the ISDN

The Ericsson systems and, it is hoped, any future systems use the ISO and CCITT recommendations wherever these have been sufficiently defined. Nevertheless, despite the importance of the ISDN to technicians in the industry, it is premature to provide here any treatment more comprehensive than a summary. *Figure 8.9a* shows the block diagram of the Ericsson Diavox Courier 1000 telephone used in the Venice trials. A comparison can be made with the purely speculative arrangement shown in *Fig. 8.6*. Where a data type device, screen, computer terminal, data store, etc., is to be connected to use the second B channel, then the terminal adaptor required is shown in *Fig. 8.9b*.

The Exchange Building

The siting of the telephone exchange is dictated by the network requirements. It is, therefore, likely that the exchange building will be located in a dense part of the exchange area where the cost of the site is at a maximum. For this reason among others, the cost of real estate and the cost of the buildings themselves represent a large proportion of the operating company investment. Earlier electromechanical systems tended to be mounted on racks of 3.5 m height and power supplies and cabling were fed from above. Exchange buildings, therefore, required a minimum 4.5 m ceiling height which is not standard

Fig. 8.9a Block diagram of ISDN telephone instrument: DIAVOX Courier 1000

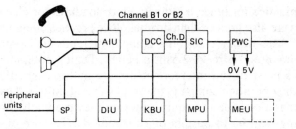

PWC	Power controller which extracts the power from the line.
SIC	S-reference controller handling layer 1 functions.
DCC	D-channel controller handling bit operations in layer 2.
AIU	Audio interface unit comprising PCM coding, decoding and tone generation.
MEU	Memory unit comprising program and data memory. The memory can be expanded for add-on features.
MPU	Microprocessor unit handling layer 2-7 functions.
KBU	Keyboard unit including interface circuits to the MPU.
DIU	Display unit comprising alphanumeric display and its interface to the MPU.
SP	Serial port for add-on facilities.

Fig. 8.9b Terminal adaptor for connecting service terminal to the ISDN

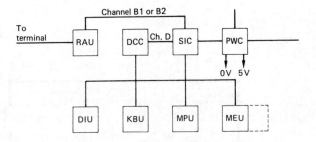

PWC	Power controller which isolates the adapter from the line and regulates the power coming from the mains.
SIC	S-reference controller handling layer 1 functions.
DCC	D-channel controller handling bit operations in layer 2.
RAU	Rate adaptation unit.
MEU	Memory unit comprising program and data memory. The memory can be expanded for add-on features.
MPU	Microprocessor unit handling layer 2-7 functions.
KBU	Keyboard unit including interface circuits to the MPU.
DIU	Display unit.

practice for office buildings. Floor loadings for electromechanical equipment were also greater than normal office practice demanded.

The newer electronic equipment tends to be smaller and lighter and is mounted on racks 2.5 m in height. Digital equipment requires far less cable as most of the interconnectors are multiplexed. It is, therefore, a relief to the operating company that replacement units will usually fit in existing buildings alongside the equipment that they gradually replace and that new buildings can be to normal office standards. The use of remote multiplexors and concentrators can also reduce the requirement for central siting of the exchange.

Because of the miniaturised nature of the new equipment and because the equipment is operating, using power, even when no calls are being processed, the power requirement has not greatly reduced over the electromechanical requirement but a new requirement has been introduced for cooling the equipment. Modern exchanges are often installed on a false floor allowing convenient access for power, cables and forced air cooling.

Digital exchange equipment is perhaps now approaching the irreducible minimum in size using known technologies. It is already at a size where the human requirements of the exchange building and the peripheral services to the exchange (power, cooling, frames) require a greater proportion of the space than the exchange equipment itself.

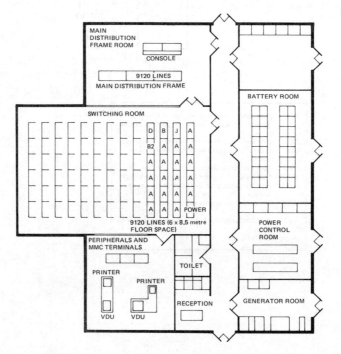

Fig. 8.10a Typical layout for a local ITT 1240 digital exchange equipped with 9120 lines; room is provided for extension to 30 000 lines

Digital Frontiers 179

These remarks are illustrated by the floor plans shown in *figs 8.10* and *8.11*. The equipment room of *fig. 8.11* occupies just 40% of the total area of the building. *Figure 8.10* supports comments made in Chapter 6 on the benefits of distributed control: the ITT 1240 installation uses 4 rack types compared to the 8 types evident in *fig. 8.11*.

Fig. 8.10*b* Typical rack layouts for ITT 1240 digital exchanges such as that in fig. 8.10a

60L	60L	POW	GS 1 & 2	POW	GS 1 & 2	30T	30T
60L	60L	GS 1 & 2	POW	GS 1 & 2		30T	30T
CONT	POW	POW	CONT	POW	GS 3	CONT	POW	GS 3	CONT	POW	POW
AIR BAFFLE			AIR BAFFLE			AIR BAFFLE			AIR BAFFLE		
60L	60L	30T	30T	32SC	32SC	30T	30T
60L	60L	30T	30T	32SC	32SC	30T	30T
CONT	POW	POW	CONT	POW	POW	CONT	POW	POW	CONT	POW	POW
TYPE A			TYPE J			TYPE D			TYPE B		

60L   60-line analog subscriber module
30T   30-line analog trunk module
32SC 32 Service circuits module
CONT Auxiliary control element
GS   Group switch
POW   Power

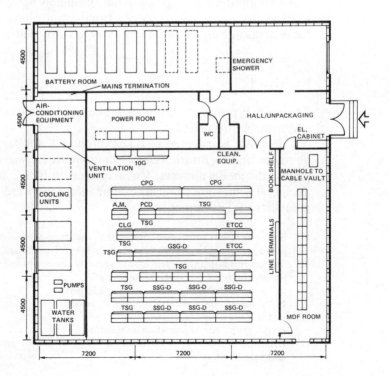

Fig. 8.11 Equipment layout: AXE 10

Chapter Summary

To complete the treatment of the digital communications switch we have moved to the periphery of the exchange. Dealing first with the function of scanning lines and trunks looking for new call attempts we have seen that the function has become integrated with the PCM coding and decoding process.

In dealing with the subscriber's line interface we outlined the problems of compatibility with the existing local network and existing subscribers' apparatus. It has been necessary to consider the new network and subscriber possibilities introduced when the digital interface is moved to the subscriber's instrument.

The only remaining interface to the trunk network was then the interface to the existing analog trunk network. While this analog network exists it will impose constraints on the extension of common channel signalling and it will require that the digital SPC exchange remains competent to deal with existing forms of channel associated signalling. We had promised in Chapter 5 to return to the subject of line and trunk testing and this is now fulfilled.

Returning to the subscriber's interface the possibilities for the integration of voice, data and other services over the one public communications network were introduced. This development is taking place now and is not yet ready for the exhaustive treatment it deserves in a textbook such as this. At this stage, we can do little more than discuss the terms and concepts in use and recommend the student to refer to current literature.

A final interface is the physical one of suitably housing the exchange equipment and providing it with services. In indicating what is required we illustrated the benefits that digital SPC technology has provided in this area.

Exercises 8

This section provides examination questions at a level equivalent to that at which the book is aimed.

The author is indebted to BTEC in the UK for permission to publish the questions and to *British Telecommunications Engineering* for permission to publish the model answers. Model answers to BTEC and SCOTEC (Scottish Technical Education Council) examination papers are published as a supplement to the BTE Journal.

8.1 State two factors which may contribute to the shape of a telephone exchange traffic graph over a period of 24 h and briefly explain the effect of each factor.

8.2 The factor controlling the provision of exchange equipment and junction line plant is (choose the one correct answer from the following):
 a) the total number of calls handled in a 24 h period
 b) the forecast busy-hour traffic at the design date
 c) the grade of service provided
 d) the size of the building which is available.

8.3 Explain the difference between the terms *concentration* and *distribution* in telephone-exchange structure.

Fig. 8.12

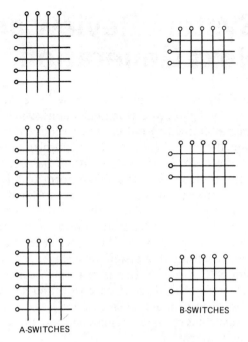

A-SWITCHES   B-SWITCHES

8.4 Explain briefly why a 4-digit local director telephone exchange can serve 10 000 lines, while a 4-digit non-director telephone exchange has a limit of about 6000 lines.

8.5 *Figure 8.12* shows a 15×15 matrix switch using three 6×4 A-switches and three 3×5 B-switches. Complete the diagram by inserting nine A–B links and three inter-aid links between the A-switches.

8.6 Draw a simple switching network to show how two incoming 4-channel digital systems can be switched to two outgoing systems by using a time-space-time digital switch.

8.7 If the traffic flow in erlangs during a specified period is known, state three items of information that can be derived from this knowledge.

8.8 The number of calls carried by a rank of exchange switches are counted at intervals of 10 min during a 1 h period. The number of simultaneous calls in progress at each count were 20, 28, 16, 23, 22, 18, 20. Find the traffic intensity.

8.9 For each of the requirements listed below, state whether in-band or out-band signalling would be used:
 a) signalling to take place during speech transmission
 b) uses standard transmission terminal equipment
 c) uses the simplest form of signal-detection circuit.

9 System Review and the Next Generation

The Shape of Systems

The object of this work has been to outline the basic principles upon which the design of any digital, SPC communications switch will be based. Further, the book has attempted to outline the various avenues that system design can follow and the undoubted dangers and problems presented by each choice. Throughout the book the Strowger system has been used to illustrate the concepts and problems inherent in any form of telephone switching. This is because the Strowger features could be explained in relatively few words and thus served as a convenient method of outlining the concepts involved. It is true to say that most of the systems prior to SPC could be comprehended in their entirety by one person, albeit after at least a year working with the system. Comprehension of a system by a single person is probably no longer possible with SPC systems. This is not only because the system has been separated into the disciplines of hardware and software, but also because the system has become more integrated and less divisible into functions. This unifying principle has operated despite our best efforts at software modularity and distributed control.

As the work progressed, digital stored program control system principles were illustrated by using examples from a few of the many systems available today. The examples were limited to a small number to enable the reader to become familiar with the concepts and terminology of these systems. It is appropriate now to look in broad outline at the many available systems and attempt to discover any unifying principles in their design.

What causes systems to differ? Manufacturer's style, the background of the design team and the network requirement of the primary customer all have an influence. The review of control architecture in Chapter 6 indicated that system size could have a profound effect on some aspects of architecture.

In discussing analog switching methods we were able to detect a rudimentary progression from direct control step-by-step towards common control link frame systems, culminating perhaps in common control overall route-choice systems. This progression was influenced and driven by the need for register control to provide urban, national and international direct dialling, and by the advent of high-speed electronic control more economically justifiable as common control. It would be satisfactory to detect similar patterns in the development of digital SPC switching.

Digital switching has not, yet, been subject to similar historical trends. From the experimental introduction of time division switches in the early 1960s there has been too short a time for the various competing system design concepts to influence succeeding designs. Designs have emerged, largely independently, having, however, much that is similar in their concepts. Without much hope of perceiving a satisfying consensus we will attempt to survey the various systems, identify their areas of difference and commonality, and postulate some of the reasons for the differences in approach.

In general, systems appear to be similar in their approach to the fundamental functions of switching, control and management (operation, maintenance, charging and network supervision). They differ, again in general, in the methods adopted to provide the less basic functions of growth, security and control message passing. These statements can be immediately challenged by pointing to systems switching in serial form or in parallel and at very different speeds and to systems using any of the control variants described in Chapter 6. For the purposes of this survey these significant variations are considered as details.

At the end of this chapter an annexe is provided detailing the main features of some 12 public exchange digital systems. This information is not exhaustive, nor is the list complete, but it is sufficient to illustrate the main sources of variety in system design approaches.

It is (just) possible to detect general differences between systems based upon the historical route taken by their development. In this respect three developmental routes are immediately apparent:

Historic   Digital SPC system developed from an earlier analog SPC system. *Examples:* AXE 10, EWSD.
Stylistic Systems developed initially for digital SPC but by design teams influenced by previous "house style" approaches. *Examples:* ESS5, LMT MT20 (uses Metaconta style), E10B, System X.
Inventive Systems developed initially for digital SPC by teams drawing inspiration from earlier projects outside telecommunications. *Examples:* System 1240.

We will consider one or more examples of each of these approaches to system design in an attempt to understand the design process.

The Historic Approach

Although the only major exporter of telecommunications equipment in Sweden, Ericsson do not have a monopoly of the home market. This monopoly is held by Televerket, a manufacturing company owned by the administration but having technology transfer arrangements with Ericsson. Ericsson, therefore, depend for their livelihood upon exports and Ericsson systems are designed primarily for world markets, not for a home administration.

Ericsson developed the AKE 13 system, an analog switching system, using reed relays as the switches and with stored program control. AXE 10 was developed initially as a digital replacement for the AKE 13 group switch using similar control hardware and employing the analog concentration stages of AKE 13. There was, at the time (late 1970s) considerable resistance among manufacturers to the concept of developing all-digital systems which might render their existing analog switches obsolescent. In Ericsson's case this resistance was overcome largely by the winning of a massive order in Saudi Arabia, a condition of which was that later tranches of the order should include fully digital switching.

The success of the AXE 10 system can be traced to its use of architecture, design concepts and software modularity techniques already proven in the AKE 13 system. Nevertheless, continuing development of AXE 10 has evidenced the need for ever more powerful processors and more extensive memory to deal with the complexity of the software. In this, as in all other digital SPC developments, initial estimates of software quantities were exceeded many times over.

The German approach to switching development has always been a cooperative one. Siemens, the major supplier to the Deutsch Bundespost (DBP) has led the cooperative developments with a lesser share being taken by the other German manufacturers. This approach was taken in developing the EWSA system of analog SPC switching until the development was abandoned in 1979. The DBP then announced that it would accept digital systems only and arranged a competitive assessment of trial systems. This competition resulted in the choice of EWSD (Siemens), a system recognisable as being out of the EWSA stable and System 1240 (Standard Electric Lorenz). Because of the DBP insistence on fully digital systems, EWSD was designed from the start as fully digital and the resulting system is a more integrated whole as a result.

The Stylistic Approach

The Bell system in the US was the first to introduce SPC analog systems and by 1982 some 30% of the Bell system was served by such systems (No. 1, No. 2 and No. 3 ESS). At the same date, the digital trunk switching system No. 4 ESS terminated 30% of the system's toll trunks. With this existing network of relatively new and efficient local exchanges there is little impetus for a rapid introduction of digital switches in the large exchanges of the urban networks. Increasing use of digital pair gain systems in the suburban and rural networks (where Nos 1, 2 and 3 ESS have not penetrated) encouraged the development of a small (by US standards) digital system No. 5 ESS.

The Bell software team has more SPC systems to its credit than any other in the world and the style of the No. 5 ESS is evidenced mainly in the software which owes much to its predecessors. Further, the American telephone system requirement is esoteric by international standards and has for long provided features which are rare elsewhere. These are all features encouraging an individual "style".

De-regulation of communications in the USA has led the Bell system to look elsewhere for sales and No. 5 ESS is now offered internationally in a CEPT 30-channel version as the No. 5 ESS-PRX in cooperation with Phillips.

The French telephone network was the least developed of all the developed countries until the early 1970s. At that time the French government took the decision to invest massively in communications to rectify the existing appalling situation. CIT Alcatel, therefore, had a very large home market for which to develop the E 10 system. With generous development funding the system was the first digital system to enter public service in production quantities and achieved notable export success. The system was developed from early trials, in which the French PTT participated, for both military and civilian switches. Being the first in the field there was no requirement for digital concentration and still today CIT Alcatel would only recommend digital concentration for those lines with a real requirement.

In the early 1970s the UK administration (then the British Post Office, now BT) received a rather bad press for standardising on two analog common control systems: TXE 2, reed switch, wired logic, for small systems, and TXE 4, reed switch, SPC, for larger exchanges. In its day, TXE 4 was remarkable for its almost perfect modularity. It was possible (and was achieved in the redevelopment for its successor TXE 4A) to completely change a module while leaving the remainder of the design untouched. This led BT to insist on a black box approach to the proposed digital SPC system, System X, and several years were spent from 1973 onwards in defining the modules in detail. This system definition made it possible for different manufacturers to develop different modules as they wished, provided of course that they met the onerous interface requirements. An important prerequisite to this strategy was the intention to use CCS as the signalling interface between modules as well as to the digital network.

System X, while using fairly pedestrian approaches to many of the digital SPC design problems, is unusual in having achieved this high degree of modularity and is the only system to use CCS for internal messages between modules.

The Inventive Approach

Prior to the UK decisions on System X there were several contenders for the processor system to be used in the new development. One of these contending systems was the Plessey PP250 processor complex whose development was generated by requirements for military communications systems and traffic control systems as well as public exchange switching. Some leading members of the PP250 development team emigrated to the US and eventually in the late 1970s congregated in the ITT Advanced Technology Centre in Connecticut.

At that time ITT switching business was booming, based largely on the Metaconta range of analog SPC systems. ITT, therefore, envisaged a rather leisurely development of digital switches based upon a toll system which would be a redesign of Metaconta (System 1220), a small local system (System 1230) and the later introduction of a very advanced system (System 1240). The Connecticut team was charged with the system definition of System 1240 and their background of distributed processing may have contributed to the quite different approach adopted to digital switching.

Market tendencies caused acceleration of the System 1240 development displacing 1220 and 1230 entirely. The French ITT company, LMT, had contributed to the early work on 1220 prior to their nationalisation so that a system rather like 1220 exists in the LMT MT20 range.

Review

Out of this wide variety of development histories has emerged a range of systems which, with one or two notable exceptions, are rather similar in architecture.

Switching has been accomplished by a multiplexor/time switch concentration stage and by a group stage consisting of separate time and space switches with the space stages in the centre and providing the growth function. Security, sometimes confined to the group switch only, is by duplication or replication of the complete network or individual switches.

Control is central but with distributed device control functions. Security of control varies between micro-synchronous duplication, hot stand-by duplication or multiprocessing as the favoured options.

Software structure is universally modular, usually employing a similar breakdown into modular functions. Languages vary though a high-level telephony-oriented language is the favoured option. It is probable that the CCITT-defined high-level language CHILL or its successors will eventually be adopted universally. There is, probably, no system which has not been forced to revert to machine code for certain heavily used functions.

Message-passing between software and hardware modules is usually by means of high-speed bus structures. All system developments have been beset by timing and volume problems in attempting to interwork complex reactions between modules by means of standard message formats. These problems have perhaps been greater for the exceptional systems such as System 1240 and System X.

The exceptions to this generality of approach are System 1240 (distributed processing, combined time and space switch module, message passing via telephony switch) and System X (message passing via CCS). AXE 10 is also exceptional in being able to spread traffic load between concentrator modules. The general pattern is also broken by those systems (UXD 5 and Protel UT 10/3 for example) which were specifically developed for small exchange application'.

The Next Generation

Engineering for this Generation

Inevitably, in treating the subject of Communications Switching at a time of rapid change, the choice of material has to be limited and cannot descend into necessary detail which may be out of date as soon as it is written. It is appropriate, therefore, to offer some words of comfort to the technician dissatisfied that subject areas necessary to daily work may not have been addressed.

The author, when he started work in Communications Switching in 1953, was able to consult a two-volume work which contained complete detail on current manual and automatic systems and considerable detail on Crossbar, Rotary and Panel systems. With a system lifetime of 40 years (UK Strowger has had a lifetime of 60 years so far) it is possible to provide such complete coverage. Our present-day systems have a design lifetime of 20 years but technological change is now so fast that it is highly unlikely that today's systems will remain in manufacture for 20 years. It is therefore unrealistic to expect comprehensive works on particular systems.

The increased rate of technological change is not the only problem facing the engineer and technician attempting to acquire a working understanding of a modern, digital, SPC switching system. The two other chief obstacles to understanding are the system complexity and, paradoxically, the system reliability. The increased complexity of modern systems has been commented upon at the start of this chapter and is evident throughout the book. The effects of increased reliability deserve further comment.

Early systems were designed on the assumption that there would be a continuous maintenance presence. At the least, the maintenance engineer was expected to routinely lubricate mechanisms, and relay sets and selectors required routine adjustment. In any case the relay logic and switching circuits were basically much less reliable than modern solid-state circuitry. The maintenance engineer therefore obtained familiarity with the system through continuous contact, a contact which was quite often with a system that was faulty. Familiarity with the system was, therefore, obtained largely without special effort. Modern systems, by contrast, are much more reliable and the maintenance engineer does not often need to find faults and never has to perform routine maintenance. Further, when a fault does occur it will usually be indicated directly by the system itself and is cured by a simple card replacement. Failing this, exceptionally, the fault will be extremely serious and probably difficult to find. The maintenance engineer will therefore be working under the pressure of a significant failure that is obscure. In such cases textbooks will be of no value. The manufacturer's handbook and, quite probably, the manufacturer's or the operating company's highly skilled trouble-shooting team will be required.

So where in all this is the comfort? Apart from the highly unusual case mentioned above, of serious and obscure failure, most of the fault maintenance is reduced to a low skill level, such as reading a visual display, perhaps conducting a few tests indicated by the display, and replacing a faulty unit. The visual display probably includes menu techniques to assist in the repair process. The off-line repair of the faulty unit is also probably assisted by computerised test equipment or the unit is returned to the factory.

This too would be cold comfort to the engineer seeing the task de-skilled and of reduced interest were it not for the other MMC features presented by the SPC system. Maintenance has been simplified but Operation and Maintenance has become possible. The maintenance engineer can now expect to operate the network in a way that was previously inconceivable.

The exchange system display not only indicates faults but also displays a wealth of information on the traffic performance of the system. This information is available, not only at the exchange but, over CCS links, at a central network management location. Furthermore, data tables containing the operating instructions for the exchange are available locally or centrally for immediate modifications with suitable safeguards against inept or malicious change. The exchanges can, therefore, be tuned to provide optimum service under current loadings. Central operations and maintenance staff can change routings, increase trunk groups, apply traffic overload sanctions, etc., as the need arises. These tuning operations are in addition to routine management tasks such as changes and additions to subscribers and their facilities, consolidation and preparation of accounts, network and exchange upgrades and additions, etc.

The result therefore of a loss of visibility of the detailed operation of the exchange, allied to a greatly simplified fault-elimination task, has been the introduction of the interest and satisfaction of a much greater network management task previously impossible to undertake.

Future Generations

We have to assume that technological pace is not going to decrease. The agenda for technological change is full already with the task of implementing the developments we have described. The digital SPC systems now beginning to go into service will accept the integrated network using CCS as it is introduced. Already we are preparing for, if not already introducing, the sophisticated processor-controlled subscriber's terminal. Though the pace of change increases, the gap between significant breakthroughs is still measured in decades. We cannot at present see the next breakthrough, which is perhaps fortunate, as we have enough to do catching up with the last two, at least.

In concert with the pace of technological change has been a movement of revolutionary change in the organisation of the industry and, particularly, of the operating companies. The two major examples of the latter reorganisation are de-regulation in the USA, splitting the Bell monolith into constituent operating, manufacturing and research entities, and privatisation and liberalisation in the UK, creating at least the possibility for multi-sourced telecommunications service and supply. It is debatable whether these organisational movements have had a common causative factor related to technology change. Certainly, the increasing similarity between data processing and exchange control techniques has influenced the trends. Certainly too, the emergence of concepts of integration of voice and data has had an effect. However the changes have come about, the emergence of competing multiple suppliers of communications service and equipment is, in turn, going to effect the directions and the pace of future technical developments.

This interplay between technology and organisational change can be illustrated with examples from recent events.

The first digital public switch offered in the USA was developed by TRW-Vidar, a newcomer to communications switching. TRW-Vidar retired from the market in 1982. The convergence of switching and transmission and switching and data processing has led newcomers to join the industry. With the reduction in manufacturing content, smaller firms can attempt manufacture with less investment. Conversely, the increased complexity of, particularly, the software content of the design has led to the need for development teams, numbered in thousands, to design a new system. The market, even though growing, particularly in the Third World, cannot support all these new entrants or even the traditional suppliers. We can expect the list of potential suppliers to contract again over the next decade, probably to a smaller number than the original contenders.

In the UK it used to be the case that only PABX equipments approved by BT could be used and BT alone could provide maintenance. Liberalisation has degraded BT to the status of a supplier and a whole new approvals organisation is in process of being established. This has resulted in an explosion of potential suppliers of PABXs and a much more difficult choice for the user between competing products, competing maintenance organisations and even competing networks.

Although the Bell ESS No. 2 SPC system introduced the concept of the PABX integrated with the public exchange, the idea did not gain popularity in the USA because it limited the customer's choice to just the one variety. Digital SPC has made the concept of the integrated PABX even more cost effective and technically attractive. The very fact that customers have been newly presented with yet wider choice will undoubtedly prevent the wide introduction of the integrated PABX for decades to come despite the technical advantages of such an introduction. Only Northern Telecom at present offer this feature.

There is a third movement affecting our industry which may be more dramatic and certainly less understood than the movements resulting from technological pace and organisational restructuring already discussed. This third movement is the impact of the telephone on the "Third World". Until relatively recently this impact of the telephone was extremely small indeed. The continuous growth of the industry over this century has been fuelled almost entirely by the demand of North Americans for the telephone, followed by a later but equivalent demand from the rest of the developed world. Today there are almost as many telephones as there are North Americans and the figures for the remainder of the developed world are approaching this (fig. 9.1). The European spurt in telephone demand occurred in the decades succeeding the Second World War and European demand is now perhaps past its peak.

Certain Third World countries, notably Latin America, have had a long familiarity with the telephone. This has not meant a high telephone penetration but rather a very different pattern of use. Telephones in Latin America were for long the property of shopkeepers and café proprietors but heavily used by the public. In most Third World countries today telephone demand far outstrips supply and the pattern of use, in so far as the demand is satisfied, is very different. It is being increasingly recognised that the telephone can be a potent tool creating the necessary advances for the poorer nations to achieve the economic and social goals that will enhance their quality of life.

Our industrial societies were built around the sources of fossil fuels and the inland waterway and railway transport networks. Our post-industrial

Fig. 9.1 Telephones per 100 population [Courtesy: AT & T *The World's Telephones*]

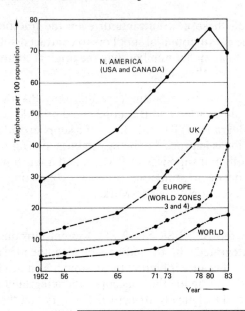

		1952	1956	1965	1971	1973	1978	1980	1983
WORLD	Phones in Service	79.4	101	182.5	272.7	312.9	423.1	472.1	486.7
	Per 100 population	3.3	3.7	5.6	7.4	8.2	14.7	16.4	17.8
	% Auto.	70.4	79.5	92.1	96.1	96.8	98.8	99.1	87.6 (?)
N. AMERICA USA and CANADA only	Phones in service	48.8	60.4	95.5	129.5	142.1	176	191.1	174.8
	% World total	61.5	59.8	52.3	47.5	45.4	41.6	40.5	35.9
	Per 100 pop.	28.7	33.1	45	57.1	61.3	73.6	77.9	69.4
	% Auto.	71.5	82.9	99	99.9	99.9	99.9	100	99.7 (?)
EUROPE (WORLD ZONES 3 and 4)	Phones in service	22.4	29.1	57.4	90.3	106.2	154.8	177.6	180.3
	% World total	28.2	28.8	31.5	33.1	33.9	36.6	37.6	37.0
	Per 100 pop.	3.6	5.2	9.2	13.7	16	20.5	23.4	39.7
	% Auto.	71.1	76.8	89.5	94.1	95.6	98.7	99.1	99.6
UNITED KINGDOM	Phones in service	5.7	6.9	10	15	17.6	23.2	26.7	28.7
	% World total	7.2	6.8	5.5	5.5	5.6	5.5	5.6	5.9
	Per 100 pop.	11.4	13.5	18.3	26.7	31.4	41.5	47.7	51
	% Auto.	72.8	77.3	91.2	98.8	99.2	100	100	100

societies are similarly influenced by modern transport networks, largely of motorways. The telephone has followed this development. In large tracts of the world today the telephone is at the leading edge of the movement of peoples out of their present agrarian economies, described by westerners as "underdeveloped". It is not fanciful to suppose that the directions of this movement will be influenced by the presence of a communications structure. Neither is it fanciful to suppose that these emergent societies will fail to be satisfied with the same features of the telephone as those to which the developed world has become accustomed.

Post-digital Communications Switching

The tendencies this chapter has discussed have set the scene for the, as yet unknown, development of our industry. While we cannot predict the developments in any detail, we have already seen the directions in which development should most properly take us.

Systems are and will be more reliable, more intimately interconnected, more able to share information, and capable of providing more information on their operation. The network using these systems will, therefore, be more ready to be delicately tuned to provide optimum service at all times. The task of the communications engineer will thus be greatly enhanced in interest and importance.

Significant penetration of telephones worldwide is greatly increasing long-distance communications traffic. This in turn creates a demand for sophisticated terminal features. The human problem of communicating over time zones, for example, is alleviated by the use of voice and text messaging features.

The change from national telecommunications administrations to competing commercial network service providers allied with the integration of voice and data demands the timely introduction of ever more rigid and complete interconnection standards. Paradoxically, the "liberalisation" tendency is making this standardisation task more difficult.

Telecommunications engineering has always been a fascinating task but one hidden from public view as a specialisation. Increasingly, as we present ever more sophisticated terminal options to an ever larger proportion of society, our task becomes more vital and more obvious to the public. The nature of the task itself is not changed by this. "Nation shall speak peace unto nation"* and our satisfying task is to assist and increase the conversation.

> "And only The Master shall praise us, and only The Master shall blame;
> And no-one shall work for the money, and no-one shall work for the fame,
> But each for the joy of the working, and each in his separate star,
> Shall draw the Thing as he sees It for the God of Things as They are!"
> Rudyard Kipling—*When Earth's Last Picture*

* M J Rendall, first governor of the BBC.

Annex to Chapter 9
Survey of Digital Switching Systems

SYSTEM:	AXE 10
MANUFACTURER:	ERICSSON
COUNTRY:	SWEDEN
NETWORK:	
ARCHITECTURE	Time concentrator. Folded TST group switch.
GROWTH	Square law growth of S stage.
SECURITY	Duplicate TST network.
CONTROL CONFIGURATION:	Duplicate synchronous (fig. 6.9a).
CAPACITY:	
EQUIVALENT LINES	200 000
BHCA	800 000
RELATIONSHIP:	Developed initially as a digital group stage for the AKE 13 SPC analog system.

SYSTEM:	DMS 100
MANUFACTURER:	NORTHERN TELECOM
COUNTRY:	CANADA
NETWORK:	
ARCHITECTURE	Space to digital concentrator 600 lines to 4 × 30 channels. TSTS folded group switch.
GROWTH	By addition of TSTS modules each switching 64 × 30 channels. Maximum 32 modules.
SECURITY	Duplicated network.
CONTROL CONFIGURATION:	Duplicate synchronous (fig. 6.9a).
CAPACITY:	
EQUIVALENT LINES	100 000
BHCA	about 250 000
RELATIONSHIP:	Design commenced as a trunk switch (DMS 200).

SYSTEM: DX 200
MANUFACTURER: NOKIA
COUNTRY: FINLAND
NETWORK:
  ARCHITECTURE   Time modules, 32, 2 Mbit/sec systems.

  GROWTH   Addition of modules.
  SECURITY   Duplication.
CONTROL CONFIGURATION: Duplicate hot standby (fig. 6.9c)
                                               N + 1 provision of peripheral processors.

CAPACITY:
  EQUIVALENT LINES   DX 210 4000; DX 220 40 000
  BHCA   DX 210 12 000; DX 220 100 000
RELATIONSHIP: Nokia development.

SYSTEM: EWSD
MANUFACTURER: SIEMENS
COUNTRY: W. GERMANY
NETWORK:
  ARCHITECTURE: Time switch concentrator.
                                                TST folded group switch.
  GROWTH   Addition to space and time stages for up to 504×8.192 Mbit/sec links.
  SECURITY   Duplicated switching network.
CONTROL CONFIGURATION: Small sizes duplicate hot stand-by (fig. 6.9c).
                                                Large sizes N + 1 multiprocessing.
CAPACITY:
  EQUIVALENT LINES   100 000
  BHCA   750 000 control capacity dependent upon choice of one of four variant control processors.
RELATIONSHIP: Development started from EWSA analog SPC system.

SYSTEM:	E10B
MANUFACTURER:	CIT-ALCATEL
COUNTRY:	FRANCE
NETWORK:	
ARCHITECTURE:	Space concentrator prior to A/D conversion or time switch concentration with CODEC per line.
	TST group switch.
GROWTH	Addition of TST modules.
SECURITY	Double dispersion of time slots in the concentration stage (URM) and between URM and group switch.
CONTROL CONFIGURATION:	Functional distribution of processors with N + 1 provision.
CAPACITY:	
EQUIVALENT LINES	60 000
BHCA	300 000
RELATIONSHIP:	Development based upon E10A system, the pioneering digital switching development.

SYSTEM:	GTD 5EAX
MANUFACTURER:	GTE
COUNTRY:	USA
NETWORK:	
ARCHITECTURE	Assumes separate line concentrator. TST folded group switch.
GROWTH	Square law growth of S stage.
SECURITY	Duplicated TST network.
CONTROL CONFIGURATION	Multiple processor pairs each duplicate hot stand-by (fig. 6.9c).
CAPACITY:	
EQUIVALENT LINES	145 000
BHCA	720 000 claimed but only 280K BHCA possible to date (1985) using 7 telephony processors.
RELATIONSHIP:	Trunk switch development later used for local switch.
	International version GTD-5C uses Protel UT 10/3 concentrator/small exchange system.

SYSTEM:	NEAX 61
MANUFACTURER:	NEC
COUNTRY:	JAPAN
NETWORK:	
ARCHITECTURE	Analog concentrator. Up to 22 duplicated TS switches folded and fully interconnected at outlets to S stage.
GROWTH	By addition of modules.
SECURITY	Duplication.
CONTROL CONFIGURATION:	Call processor pair per 1 or 2 network modules. O & M processor pair per exchange. Each pair operate in "synchronous stand-by" (fig. 6.9a and c).
CAPACITY:	
EQUIVALENT LINES	100 000
BHCA	700 000
RELATIONSHIP:	

SYSTEM:	PROTEL UT 10/3/GTD 5C
MANUFACTURER:	ITALCOM Joint venture: Italtel, GTE, Telettra.
COUNTRY:	ITALY
NETWORK:	
ARCHITECTURE	UT 10/3 is small exchange part of family. Main exchange GTD 5C similar to GTD 5 EAX. Up to 16 T modules fully interconnected.
GROWTH	Addition of modules.
CONTROL CONFIGURATION:	For UT 10/3 module control processor per module plus central administration processor. Duplicate hot stand-by (fig. 6.9c).
CAPACITY:	
EQUIVALENT LINES	20 000
BHCA	120 000
RELATIONSHIP:	UT 10/3 was Telettra development.

SYSTEM:	SYSTEM X
MANUFACTURER:	PLESSEY/GEC
COUNTRY:	UK
NETWORK:	
ARCHITECTURE	Digital concentrator plus TST group switch.
GROWTH	Addition of T stages; square law growth of S stage.
SECURITY	Duplicated network.
CONTROL CONFIGURATION:	Multiprocessing (fig. 6.9e).
CAPACITY:	
EQUIVALENT LINES	60 000
BHCA	500 000*
RELATIONSHIP:	Fundamental philosophy of separate sub-systems. A version of #7 CCS is used to communicate between subsystems.

SYSTEM:	SYSTEM 12
MANUFACTURER:	ITT
COUNTRY:	BELGIUM, W. GERMANY, ITALY, SPAIN and USA.
NETWORK:	
ARCHITECTURE	Distributed architecture based on self-seeking switch module combining T & S switching.
GROWTH	Addition of T/S stages.
SECURITY	Duplication of network paths.
CONTROL CONFIGURATION:	Distributed control (fig. 6.9f).
CAPACITY:	
EQUIVALENT LINES	100 000
BHCA	In theory, unlimited. Capacity increased by adding control elements.
RELATIONSHIP:	Basic idea of distributed processing probably emerged from Plessey development of PP 250 processor in early 1970s. Some of Plessey team formed nucleus of ITT 1240 system definition team.

---

* UK throughput figures include allowance for overload situations. They may therefore appear conservative compared with claims of other systems.

SYSTEM: UXD 5
MANUFACTURER: Designed by BT, manufactured by Plessey and GEC.
COUNTRY: UK
NETWORK:
   ARCHITECTURE   96-port and 32-port multiplexes with T-stage interconnection.
   GROWTH   Addition of multiplexes.
   SECURITY   Duplicate T stage.
CONTROL CONFIGURATION: Duplicate cold stand-by (fig. 6.9b).
CAPACITY:
   EQUIVALENT LINES   600
   BHCA
RELATIONSHIP: Developed from the Monarch digital PABX.

SYSTEM: 5 ESS
MANUFACTURER: AT & T
COUNTRY: USA
NETWORK:
   ARCHITECTURE   Analog concentrator.
                                 TST group switch folded at S stage.
   GROWTH   By addition of switching modules.
   SECURITY   Duplication.
CONTROL CONFIGURATION: Duplicate hot stand-by (fig. 6.9c).
CAPACITY:
   EQUIVALENT LINES   60 000
   BHCA   150 000
RELATIONSHIP: Similar control concepts and software practice to other ESS products.

Answers to Exercises

1.1 20 subscribers can call at once.
1 AB link in use blocks calls between 4 other terminals at each side of the network.

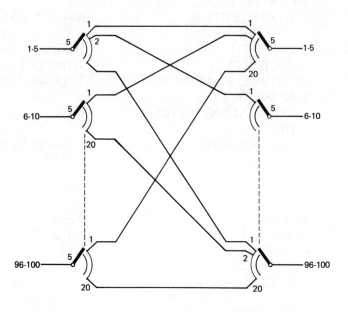

1.2
M1 operating   Marks with an earth all A switch relays in row y.
M2 operating   Marks with −50 V all A switch relays in column x. Only relay Axy is backed by an earth from M1 and this relay operates.
Axy operating   Extends the M1 marking earth to all B switch relays in row q.
M3 operating   Marks with −50 V all B switch relays in column p. Relay Bpq does not operate because it is short circuited by −50 V from M2.
M2 releasing   Removes the short circuit from relay Bpq which operates to earth via M1, Axy, Bpq, M3 to battery.
Bpq operating   etc.

1.3 12 B switches each with 12 inlets.
A–B link loading is 0.13 E.

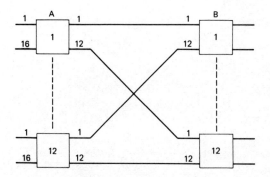

Answers to Exercises

1.4

Advantages	PAM TDM	PCM TDM
Multiplexing	✓	✓
Simple terminals	✓	—
Low bandwidth	less anyway	no
Simple regeneration	—	✓
Noise immune	—	✓
Low distortion	—	✓
Voice/data compatibility	—	✓

1.5

Send proceed to send	Pre-selection
Recognise clear forward	Release
Start charging	Conversation
Mark connection path	Call completion

1.6

End of sending	Routing
Backward clear	Line
Dial tone	User
Meter pulse	Line
Seize pulse	Line
Inter-digit pause	Routing

2.1 Figure 2.1
a) Earth unbalance
Interference from radio, power, etc.
b) Echo
c) Echo
Go/return crosstalk

2.2

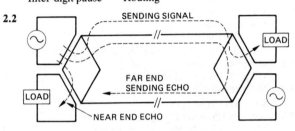

2.3

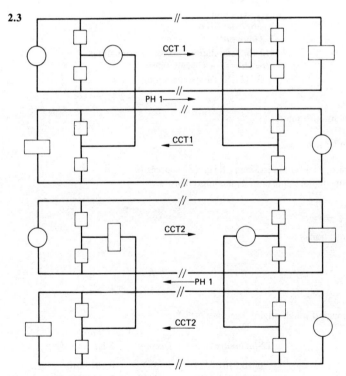

2.4 Frequencies in the Go and Return directions are separated. The directional filters allow the A–B and B–A paths to be separated after both directions have passed through the common amplifier.

2.5 The outband signal frequency must be sufficiently far from the top of the speech band (to avoid voice imitation) and also far from the channel cut-off frequency (to avoid signal limiting by the band-pass filter).

2.6 Crosstalk is recognisable noise, speech, data or switch noise. Transmission noise introduced by the transmission circuit active devices tends to be of uniform spectrum (white noise).

3.1 64

3.2 Frame alignment allows the receiver to sort out which channel is which.

3.3 See figure on page 201.

3.4
Frequency range	10–33.3 Hz
Sampling rate	66.67 Hz
Bits per sample	4
Time slots per frame	5
Channels per frame	4
Output bit rate	1333.3 bits/sec
Encoding law	Linear
Signalling capacity	133.33 bits/sec

3.5 Sampling frequency 200 Hz
Output bit rate 14.4 kHz

3.6 Simpler signalling possible.
Common channel possible.

3.7 Matches US rural FDM system modularity.
Used similar technology and components.

3.8 Need for signalling channel.
Need to provide nationwide US service.

3.9 Different quantising law.
No additional alignment bit in UK system.
Signalling in multi-frame.
No bit stealing in UK system.

3.10 a) *Bell D2*. Frame alignment pattern spread over 12 frames. System has to study up to 193 bits, each for 12 frames, to detect pattern.
Time to recognise is therefore on average:
$$96.5 \times 12 \times 125 \times 10^{-6} = 0.14 \text{ seconds.}$$

b) *CEPT*
Frame alignment recognised in
$$\frac{256}{2} \times 2 \times 125 \times 10^{-6} = 0.032 \text{ seconds.}$$

Multi-frame alignment recognised in
$$0.032 \times 16 = 0.512 \text{ seconds.}$$

3.11 24/32 word
24/30 channel
1.536/2.048 MHz
7/8 bit word
4/2 kbit/sec channel associated signalling.
–/64 kbit/sec CCS capability.
–/G732 CCITT specification.

3.12 Zero crossings indicate frequency.
Frame alignment pattern indicates start of frame.
Multi-frame alignment pattern indicates start of multi-frame.

3.13 4.096 MHz

3.14 No answer given. Resulting diagram will be similar to *fig. 3.14*.

4.1 Provided the digital system is correctly dimensioned and commissioned, the analog sources of error will either have no effect or will make the digital signal entirely incomprehensible.

4.2 Synchronisation
Slip
Quantisation distortion.

4.3 Provides continuous "background" to communications.
Is particularly apparent with "quiet talkers".

4.4 See figure on page 201.

4.5 Not amenable to digital processing.
Not easily approximated by linear sections.
Table values not easily stored in binary form.
Insufficient companding function (not steep enough through origin).

4.6

12-bit binary	Decimal	μ-law coding 8-bit
0010,0101,0111	599	0,100,0011
0010,1011,1000	696	0,100,0110
−0000,0010,1010	−42	1,001,0010
0001,0111,0000	368	0,011,0101
−0110,0000,1100	−1548	1,101,1000

4.7

12-bit binary	Decimal	A-law coding 8-bit
0010,0101,0111	599	0,101,0010
0010,1011,1000	696	0,101,0101
−0000,0010,1010	−42	1,001,0100
0001,0111,0000	368	0,100,0101
−0110,0000,1100	−1548	1,110,1000

Answers to Exercises 201

3.3

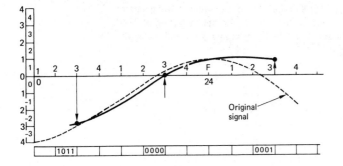

Channel 2 received, demultiplexed, decoded, integrated

4.4

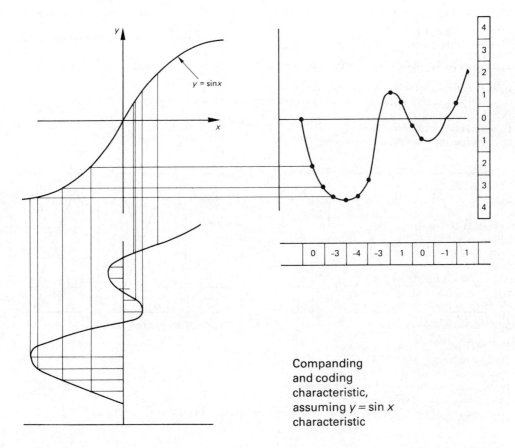

Companding and coding characteristic, assuming $y = \sin x$ characteristic

202 Introduction to Digital Communications Switching

4.8 Since the sampling rate is 8 kHz a sequence of 8 samples repeated cyclically will generate a 1 kHz sine wave. For convenience, we choose the samples 45° apart starting at 22.5° so that we only need to determine 2 values. For a half power sine wave and μ law the maximum amplitude is

$$0.707 \times 8159 = 5768$$

The values we want are therefore

| 5768 | $\sin(22.5°) = 2207$ at 22.5°, 112.5°, etc. |
| 5768 | $\sin(67.5°) = 5329$ at 67.5°, 157.5°, etc. |

These encode as shown.

Angle	Decimal	13-bit binary	8-bit μ-law
22.5°	2207	0,1000,1001,1111	0,110,0001
67.5°	5329	1,0100,1101,0001	0,111,0100
112.5°	2207	as 22.5	0,110,0001
157.5°	5329	as 67.5	0,111,0100
202.5°	−2207	etc.	1,110,0001
247.5°	−5329	etc.	1,111,0100
292.5°	−2207	etc.	1,110,0001
337.5°	−5329	etc.	1,111,0100

For A-law the maximum value is

$$0.707 \times 4096 = 2896$$

and the resulting A-law codes, excluding the sign bit are

| 22.5° | 110,0001 |
| 67.5° | 111,0100 |

The student is left to fill in the gaps of the A-law solution.

4.9

A-law	Value	μ-law
0,010,1001	102	0,001,1001
1,100,1101	−472	1,011,1101
1,111,0000	−2112	1,110,0000
1,010,1001	−102	1,001,1001
0,111,1111	4032	0,110,1111
1,001,0101	−43	1,000,0101
0,010,0111	94	0,001,0111

4.10 Delta modulation with perhaps adaptive gain control and with methods to encode silent periods.

4.11 Slope overload and granular noise.

4.12 2 kHz

4.13 566 μsec

5.1 Early PCM transmission was in the (local) trunk network. It is easier to develop a TDM switch for concentrated ready coded traffic.

5.2 PCM LIC complexity.
PCM TDM switch efficiency.
Switch and control use similar protocols in PCM TDM.

5.3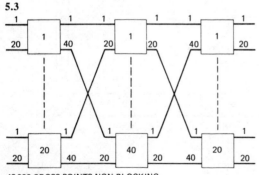

48 000 CROSS POINTS NON-BLOCKING

5.4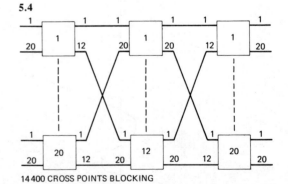

14 400 CROSS POINTS BLOCKING

5.5 For 512 inlets $n = 16$ $k = 7$
For 2048 inlets $n = 32$ $k = 10$

5.6 Time slot delay is $125/32$ μsec per time-slot.
In the go direction there are $11 - 3 = 8$ time-slots.
In the return direction there are $(32 - 11) + 3 = 24$ time-slots.
Assuming no delay in the receiver then the total round trip delay is $(8 + 24) \times \dfrac{125}{32} = 125$ μsec

Answers to Exercises 203

5.7   PCM TDM is more easy to manipulate than analog samples.
5.9   A multiplicity of time/space interfaces is undesirable and expensive. (It is assumed the student has produced a diagram similar to *fig. 5.17*.)
5.11 End-to-end route choice requires an ideal compromise between switching network simplicity and control complexity.
5.14 2048 subscribers have onward access via 11×32 channels.
Concentration ratio is 2048-to-352 or 5.82-to-1.
The remote LSM will employ channel 0 of every ETC and channel 16 of two of them.
There are therefore available for traffic 9×31 channels and 2×30 channels.
Concentration ratio is therefore 2048 to 339 or 6.04 to 1.

5.8

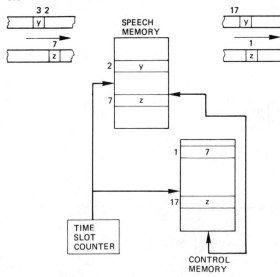

5.15 To replace working LIC with spare LIC operate relay A and relay B_S.

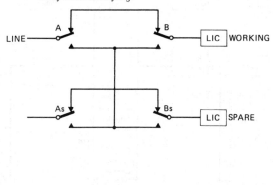

6.1   In real time.
Must always be operational.
Is called into operation by random events.
6.2   *Figure 6.3* shows a 200-line group selector. HA switches to one set of wipers with access to 100 lines; HB switches to a second set.

6.3 & 4

National and International	$0------$			
Service	$1--$			
Exchange A	20 (1–6) × × ×	Exchange D	24 (1–4) × × ×	
Exchange B	21 (1–3) × × ×	Exchange E	24 (7 & 8) × × ×	
Exchange C	22 (1–7) × × ×	Exchange F	25 (1–8) × × ×	
Adjacent Exchange	3 AB × × ×	Exchange G	26 (1–5) × × ×	
	if A–G can translate			
	or 23 (1–0) × × ×			
	no translation			

6.6

Markers	Identify and connect switch paths.
Originating registers	Store DN and convert to routing information.
Incoming registers	Store incoming information and convert to routing/terminating information.
Register access switch	Register access provision.
Register junctor	Interface of register to network.

6.8   *a)* Duplicate synchronous, hot stand-by.
b) Dual load sharing, multi-processing, distributed control.

6.9 The figure below, a processor cluster block diagram for System X, gives detail in answer to this question.

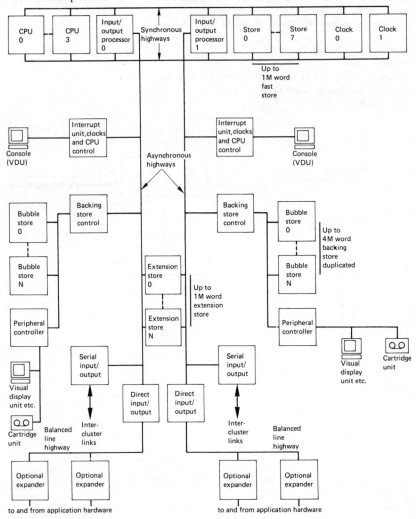

7.3 4.096 MHz to cater for the HDB3 frequency-doubling feature.

7.4
4.096 MHz	HDB3 coding
2.048 MHz	CEPT 30-channel bit timing
256 kHz	CEPT 30-channel octet timing
512 kHz	(possibly) to identify quartets in channel 16. This could be performed just as well with a 4-bit counter delay on the 256 kHz clock.
64 kHz	CEPT 30-channel frame
4 kHz	(possibly) CCITT 30-channel multi-frame.

7.7 $\dfrac{2}{3} \times \dfrac{10^{-10} \times 1.544 \times 10^6 \times 3600 \times 24}{8 \times 24} = 0.046$ slips per day.

7.10 Minimum delay is the duration of the sent signal within the limits

10 frames ⩽ signal ⩽ 30 frames

7.12 $\dfrac{1500 \times 0.85}{2 \times 0.04} \times 2 = 15\,938$

Note: Twice originating traffic per subscriber to obtain total traffic per subscriber. Twice the number of subscribers so obtained as there are two ends to each connection.

7.13
A–F	no CCS
A–B	associated
D–network	quasi-associated
C–A	non-associated
B–D	associated
D–C	quasi-associated

7.18 $\dfrac{64\,000}{32} = 2000$ LSU per second

Signal information capacity is $2000 \times 4 = 8000$ bits per second. This is because there are 4 bits of signal information in the 28-bit LSU. It is assumed that 4 stuffing bits are added to each LSU.

7.19 Let the messages consist of an average of N octets.

$$\text{Message rate without 1 insertion} = \dfrac{64\,000}{8(N+1)} \text{ messages/second}$$

$$\text{Message rate with 1 insertion} = \dfrac{64\,000}{1 + 8(N+1)}$$

Hence, loss in efficiency due to 1 insertion in flag $= \dfrac{8(N+1)}{1 + 8(N+1)}$

7.20 MSU length in octets:

CK	2
SIF	2
SIO	1
LI	1
FSN/B	1
BSN/B	1
F	1
	9 octets

Message rate $= \dfrac{64\,000}{9 \times 8} \times 2 \times 8 = 14.222$ kbit/sec.

8.1 a) The type of customer served by the exchange; that is, business and residential. Business customers tend to generate mid-morning and mid-afternoon traffic peaks, while residential customers are mainly responsible for evening peaks where advantage is taken of cheap-rate tariffs.

b) The location of the exchange. For example, an exchange serving an area of specific industry such as a fishing community may well have a traffic peak in the early hours of the morning because of traffic to and from fish markets.

8.2 b) The forecast busy-hour traffic at the design date.

8.3 A typical telephone exchange serves several thousand customers and it would be highly uneconomical to provide each customer with individual routing equipment because of its complexity and cost. Also, since only a limited number of the customers connected to an exchange will be making simultaneous calls, it is more economical to *concentrate* the traffic from all customers to a limited amount of routing equipment. After a call has been routed through the exchange, it is necessary to *distribute* the calls to the desired line by expanding the limited number of routes to give access to all the exchange customers.

8.4 In a director telephone exchange, access to STD, outgoing local junctions and services is provided from first code selector levels. The numerical selectors can therefore provide access to local customers off all their levels, allowing the full numbering range of 0000–9999 to be used. In a non-director telephone exchange, because all access is provided from the first group selector levels, certain levels must be reserved for STD access (level 0), access to outgoing junctions (levels 8 and 9) and service calls (level 1). Thus only levels 2–7 are available for local subscriber access, allowing only the numbers 2000–7999 to be used.

Answers to **8.5** and **8.6** are given on page 206.

8.7 a) The average number of simultaneous calls during the specified period.

b) The portion of the specified period for which a circuit is engaged.

c) The number of calls which originate during a period equal to the average holding time of the calls occurring in the specified period.

8.8 The traffic intensity is given by the average number of simultaneous calls in progress during the 1 h period.

The average number of simultaneous calls in progress is given by

$$\dfrac{20 + 28 + 16 + 23 + 22 + 18 + 20}{7} = \dfrac{147}{7} = 21.$$

Therefore, the traffic intensity $= 21$ erlangs.

8.9 a) Out-band
b) In-band
c) Out-band.

206 Introduction to Digital Communications Switching

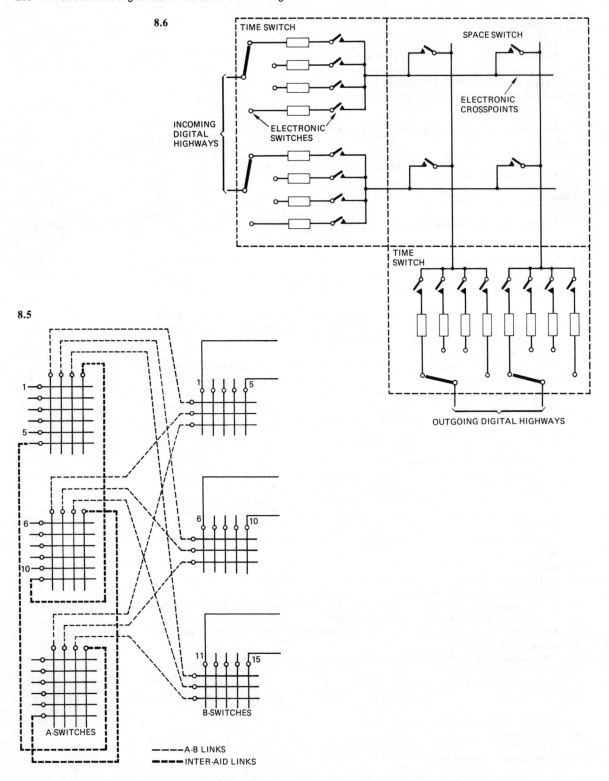

Glossary

AC (signalling) uses alternating current signals usually in or near the voice frequency band (0–4 kHz).

A-law The coding law used by the CEPT 30-channel PCM system.

ASU Acknowledgment signal unit. CCITT #6 terminology.

AT&T The American Telephone and Telegraph Company. AT&T is the manufacturing arm of the Bell organisation. Bell Laboratories provide development and, until recently, the Bell operating companies were the wholly owned chief customers.

Availability A measure of the number of individual circuits available to a particular inlet to the switching stage, e.g. a 12×16 coordinate array has an availability of 16.

BHCA Busy hour call attempts. A measuring unit to dimension SPC exchange call processing power.

BORS(C)HT Mnemonic describing the functions of a subscriber's exchange line interface. Battery, Overvoltage, Ringing, Supervision, (Coding), Hybrid, Test. The C is omitted in analog environments.

BPO British Post Office.

BSN Backward sequence number. CCITT #7 terminology. BSN = value of FSN of last SU correctly received.

CCITT Comité Consultatif International Télégraphique et Téléphonique. International Telegraph and Telephone Consultative Committee. The telephone and telegraph standards-setting body of the International Telecommunications Union (ITU).

CCS Common channel signalling. Any form of signalling system where the signals related to a number of traffic circuits are sent via a separate common channel.

CEPT Comite European Postes et Telecommunications.

CHILL CCITT High-level Language. A telephony-oriented high-level language increasingly employed for switching system software. CCITT Recommendation Z 200, Fascicle V1.8.

Co-axial Cable consists of a centre conductor surrounded by a second conductor in the form of a tube co-axial with the centre conductor.

Companding The process of a variable *compression* of the amplitude of signals at different frequencies before transmission and *expansion* after receipt in order to reduce the effect of noise.

CRE Call reference equivalent.

Crosstalk Interference which is recognisable as intelligible voice or data.

DASS Digital access signalling system. The CCS for use on links between private networks and the PSTN. CCIT Recommendation Q 920–931.

dB Decibel. A unit to express the ratio between two amounts of power.

$$\mathrm{dB} = 10 \log_{10} \frac{P_1}{P_2} \quad \text{hence} \quad \mathrm{dB} = 20 \log_{10} \frac{V_1}{V_2}$$
$$= 20 \log_{10} \frac{I_1}{I_2}$$

DC (signalling) using signalling codes consisting of varying-length direct-current impulses.

DPNSS Digital private network signalling system. The UK CCS for use on links between private exchanges.

DUP Data users part. CCITT #7 terminology. A particular realisation of the #7 level 4 protocols for use n data switching and transmission.

Duplex Transmission is bi-directional. Information can be passed in both directions at once.

EFS Error-free one-second intervals. A measure of performance of data transmission subject to synchronisation slip distortion.

Erlang Unit of communications traffic. Defined as the average number of calls existing simultaneously.

F Flag. CCITT #7 terminology. Defines the start of a message.

FDM Frequency division multiplex.

FIB Forward indicator bit. CCITT #7 terminology. FIB inverted by transmit end when a re-transmission is commenced.

FISU Fill-in signal unit. CCITT #7 terminology. Signal unit sent when no other information is available for sending.

FMM Finite message machine. A software (and hardware or virtual machine) module which performs a defined function on receipt of a defined message, the function ending with the sending of a defined message to another FMM.

FSN Forward sequence number. CCITT #7 terminology. Value incremented by 1 each time a new message SU is sent.

GOS Grade of service. Measure of traffic efficiency expressed as "1 lost call in 100" or 0.01. (See also **Loss**.)

GSC Group switching centre. The name given in the UK to the telephone exchange that forms the entry point to the national subscriber dialling network. The GSC holds the register translator equipment to provide the originating routing for national and international calls.

Hardware The electrical circuits and assemblies of circuits whose logical operation is specified by software.

HDB3 High-density bipolar modulus 3. A line coding protocol used on both the D2 24-channel and CEPT 30-channel PCM transmission systems.

Header The first portion of a message label indicating the start of a new message.

HF Radio. High-frequency radio in the frequency range 5–50 MHz.

Hz Hertz. Unit of measurement of frequency. 1 Hertz = 1 cycle per second.

IAM Initial address message. CCITT #6 terminology.

IDA Integrated digital access. The name applied to the first UK application of the ISDN.

IDN Integrated digital network.

ISDN Integrated services digital network.

ISO International Standards Organisation.

ISU Initial signal unit (of a MUM). CCITT #6 terminology.

In-band Signalling within the voice frequency band of 300–3400 Hz.

ITU International Telecommunications Union.

Jack The name used for the plug portion of a manual board plug and socket.

Junction The term used in the UK to describe a local trunk between two local exchanges.

Junctor A link between the outlet of a switch matrix and the inlet of the succeeding matrix.

kbit/sec Kilobit per second. Unit of measure of data transfer rate. 1 kbit/sec = 1000 bits per second.

LAN Local area network. Used to describe communications network facilities between cooperating computers and terminals.

Label The portion of a message, usually a data message, which indicates its nature and type.

LI Length indicator. CCITT #7 terminology. Value of LI indicates the number of octets in an SU following the LI and preceding the check bits.
LI = 0 FISU
LI = 1 or 2 LSSU
LI = 2 to 62 MSU
LI = 63 MSU with 62–272 octets.

LIC Line interface circuit.

Loss 1 Traffic loss. A measure of calls lost due to congestion. Often expressed as Grade of Service, e.g. 1 lost call in 100.
2 Transmission loss. Loss due to impedance of transmission medium and circuits. Expressed in decibels (dB) with respect to a reference.

LSI Large-scale integration.

LSSU Link status signal unit. CCITT #7 terminology. Signal unit sent, in the absence of messages, containing information on the status of the signalling link.

LSU Lone signal unit. CCITT #6 terminology.

MMC Man/machine communication.

Modulation The process of modifying the nature of a signal by using a different signal mixed with it in accordance with a defined rule.

MTP Message transfer part. CCITT #7 terminology. Comprising: Level 1, Signalling data link; Level 2, Link control; Level 3, Common transfer functions.

MTS Message transmission system. The UK version of CCITT #7 used between and within System X exchanges.

Multiple An arrangement of wiring to connect many appearances (on a switch outlet or manual board plug socket) to a single circuit. A circuit consists, typically, of two "speech" wires and a control wire.

Multiplex A combination of many information channels on to a single transmission bearer.

MUM Multi-unit message. CCITT #6 terminology.

OSI Open Systems Interconnection.

Out-of-band (or outband) Signalling which uses signal frequencies within the voice channel bandwidth but outside the voice channel, e.g. signals of 0–300 Hz or of 3400–4000 Hz.

PABX Private Automatic Branch Exchange. A telephone switch in private ownership connecting extension telephones to each other and to the public exchange via exchange lines.

Patch A temporary change applied to make a software program work correctly. The patch may not obey the full rules employed when writing new programs or properly modifying existing programs.

PCM Pulse code modulation.

Phantom A circuit formed by using the two-wire loop of another simplex channel as one leg of a further channel. The second leg is formed in a similar fashion from another simplex channel.

Protocol The convention used to communicate over a data or other link, e.g. "Dear Sir—Yours faithfully" is a protocol for a communication between strangers.

PSTN Public Switched Telephone Network.

RAM Random access memory.

ROM Read-only memory.

SIF Signalling information field. CCITT #7 terminology. The body of a #7 MSU.

Simplex Simplex transmission is uni-directional only. Information is passed in one direction from sender to receiver.

SIO Service information octet. CCITT #7 terminology. SIO defines the user part and whether the service is national or international.

Software The programs written and stored electrically in memory that control the hardware of the control processors and hence the exchange.

Space division (Switching). Any form of switching system which allots a single physical circuit to each connection.

SPC Stored program control.

SSU Subsequent signal unit (of a MUM). CCITT #6 terminology.

STD Subscriber trunk dialling. The name given to the national subscriber dialling facility in the UK.

Step-by-step (system) The common name given to the Strowger system of two-motion selector switching.

Strowger Almon B. Kansas City undertaker who invented the step-by-step switching system and the rotary dial.

SU Signal unit. CCITT #6 and #7 terminology.

SYU Synchronisation signal unit. CCITT #6 terminology.

TDM Time division multiplex.

Tropospheric scatter radio systems Provide directional, over-the-horizon, radio links in the VHF, UHF and SHF bands (100 MHz to 10 GHz) by detecting signals reflected from the troposphere using large, sensitive antennae.

Trunk A speech circuit between two exchanges.

TUP Telephone users part. CCITT #7 terminology. A particular realisation of the #7 level 4 protocols for use in telephone switching.

TXE 2 UK small local exchange system using electronic common control of a reed relay matrix switch.

TXE 4 UK large local exchange system using stored program control of a reed relay matrix switch. Probably the first example of distributed stored program control. A later version, TXE 4A, uses increased software control and enhanced MMC features.

μ-law The coding law used by the US standard 24-channel PCM systems.

Virtual machine A composite module containing both hardware and software which appears to the remainder of the software as a single functional entity.

Index

AC, alternating current, signalling 22
Adaptive gain control 58
AKE 13 system 183
A-law PCM coding 44, 57
   table 56
Aliasing 62
Aligner mechanism 66
Alignment, frame 34
Analog signalling 22, 135, 136
Analog space division requirement 77
Application layer, OSI 162
Architecture, of switching 86
Array architecture, TDM switching 84
Associated signalling 143
Availability 9, 105
AXE system 90, 108, 116, 179, 183, 192
AXE 10, APZ 210 processor 121

Basic group 22
Basic message 120
BHCA load 110
Blocking 9
Blocking arrays 74
BORSCHT 90, 170
Building, exchange 176
Burst mode 168

Cable fill 63
Call completion 13, 113
Call control 13
Call progress 113
Calling rate 4
Carrier frequency 20
Carrier telephony 20, 22
Carrier telephony, wire carrier 20
Central battery system 135
CCITT Recommendation
   G121 26
   G711 44, 55, 58, 68
   G732 38, 44
   G733 38, 42, 44
   Q251 146
   Q400–480 139
   Q421–424 (R2) 140
CCITT Signalling System R2 139
CCITT Signalling System No 6 143, 145, 146
CCITT Signalling System No 7 145, 150
CCS (common channel signalling) 16, 42, 95, 136, 141
   advantages/disadvantages 142
CEPT PCM system 38, 44
Channel 16 signalling 44
Channel 16 signalling extraction 140
Channel associated signalling 16, 140
Channel capacity, of a time switch 79
Channel modularity, in switching 95

Charging 14
Circuit switching 5
Clear signal protocol 138
Clos, condition for non-blocking 74
Co-axial cable systems 22
Coding comparison 57
Coding laws 44
Coding
   non-uniform 52
   PCM 33
   uniform 50
   using diode compandor 52
Cold stand-by control 108
Common channel signalling (CCS) 16, 42, 95, 136, 141
Common control 10
Common service distribution 172
Communications switch, objective 3
Companding 25, 39, 52
Companding, logarithmic 54
Concentration, PCM TDM 89
Concentration switching 86
Concentrators, remote 94
Control, difference between computer and exchange 99
Control architectures 107 et seq
Control functions 106
Control memory 79, 92
Control, of exchange 17
Control, security of 106
Control software 112
Conversation phase of call 13
Co-ordinate crosspoint switching systems 8
Cost distribution of exchanges 71
CRE (call reference equivalent) 26
Crosstalk 23, 25, 62
Cyclic re-transmission 155

D1 PCM system 38, 39
D2 PCM system 38, 40
DASS (digital access signalling system) 145, 173, 176
Data link layer, OSI 162
Data
   permanent 115
   semi-permanent 115
   software 113
   temporary 115
Data user part (DUP) 151
DC, direct current, signalling 22
Delay, of TDM switching 85
Delay, timing delay of PCM transmission 64
Delta modulation 60
De-regulation 188
Despotic synchronisation 66, 130, 134

Device handlers 119
Differential PCM 59
Digital access signalling system (DASS) 145, 173, 176
Digital hybrid 168
Digital private network signalling system (DPNSS) 145
Digital signalling 137
Dimensioning, of step-by-step exchange 101
Directed message 120
Director system 15, 104
Distortion 25
Distributed control 111
Distributed control, step-by-step 101–104
DMS 100 system 192
Double ended control, of synchronisation 133
DPNSS (digital private network signalling system) 145
Dual load sharing control 109
DUP (data user part) 151
Duplex transmission 20
Duplicate cold stand-by control 108
Duplicate hot stand-by control 108
Duplicate synchronous control 107
DX 200 system 193

EFS (error-free 1 sec intervals) 131
Empress exchange 70
Entraide 88
Erlang 5
Error control
   #6 149
   #7 154
Error detection 155
Error-free 1 sec intervals (EFS) 131
E 10 systems 89, 184, 194
Events, telephonic 113
EWSA system 184
EWSD system 184, 193
Exchange building 176

FDM 1, 7
Filtration 60
Final selector 101
Finite message machine (FMM) 120
Fold-over distortion 62
Frame alignment 34
Frequency division 7
Frequency division multiplex (FDM) 20
Functional division of software 119

Grade of service 5, 9
Group selector 100
Group switching centre (GSC) 17
GTD 5 EAX system 194

Index

HDB3 line code *47, 62, 154*
Header *6*
HF (high-frequency) radio links *22*
Hot stand-by control *108*

IDA (integrated digital access) *176*
Idle channel noise *57*
IDN (integrated digital network) *175*
Impedance, transmission *23*
Implementation complexity of TDM switching *84*
In-band signalling *22*
Input associated control, of time switch *80*
Integrated digital access (IDA) *176*
Integrated digital network (IDN) *175*
Integrated networks *175*
Integrated services digital network (ISDN) *175*
Interaide *88*
Interference *62*
Inter-processor link *108*
Interrogation *14*
Invention of telephone *1*
ISDN (integrated services digital network) *175*
ITT Metaconta system *109*
ITT Pentaconta system *88*
ITT System 1220 *185*
ITT System 1230 *185*
ITT System 1240 *112, 119, 178*

Kilostream *60*

Label *6*
Laboritaire Central Telephonique *2*
Liberalisation *188*
LIC (line interface circuit) *170*
Line and trunk testing *175*
Line coding
  PCM *46*
  HDB3 *47*
Line interface circuit (LIC) *170*
Line signals *14, 15*
  D1/D2 PCM *42*
Line signal extraction *140*
Line testing *93, 175*
Link frame switching *88, 105*
LMT MT20 system *185*
Loading coil *1, 25, 38*
Load sharing control *109*
Local multiplex *172*
Loop-disconnect signalling *14, 15*
Loss, traffic *5*
Loss, transmission *23, 26*
Low bit rate channels *173*
LSI *2*

Maintenance *187*
Marker central systems *105*
Marking *18*
Mean bit error rate *149*

Message switching *5*
Message transmission system (MTS) *145*
Metaconta system *109*
(SS) MF4 (push-button) signalling *139*
Microwave radio *1, 22*
Mid-riser coding *50, 57*
Mid-tread coding *50, 56*
Modularity, of software *114*
Modulation *7*
Moorgate exchange *70*
Morse code *1*
MT range of digital switches *110*
MTS (message transmission system) *145*
Mu-law PCM coding *42, 54*
Multi-drop *172*
Multi-frame *38*
Multiple unit message (#6) *148*
Multiplex, local *172*
Multiplexing *1, 7, 20, 34*
Multi-processing control *110*
Mutual synchronisation *66, 130*

National reference clock *66*
NEAX 61 system *195*
Network layer, OSI *162*
Network synchronisation *128*
Noise *23, 25*
Non-blocking switch arrays *74*
  minimum condition *74*
Non-director exchange *100*
Non-linear quantisation *39*
Non-uniform coding *52*
NTSC *57*
No 1 ESS *184*
No 2 ESS *184, 189*
No 3 ESS *184*
No 4 ESS *81 et seq, 184*
No 5 ESS *184*
No 5 ESS-PRX *184, 197*
Numbering systems *101*
Nyquist's sampling theory *12, 62*

Open system interconnection (OSI) *160*
Operating systems *120*
Operation and maintenance *187*
OSI (open system interconnection) *160*
Out (of) band signalling *22*
Output associated control *80*
Overall route choice *105*
Overload control *124*

PAL *57*
PAM *12*
Panel system *104*
PCM *2, 29*
PCM, example system *29*
PCM patent *19*
PCM, practical systems history *37*
PCM switching *70*
Pentaconta system *88*
Permanent data *115*
Phantom *20*

Physical layer, OSI *162*
Plesiochronous operation *67, 130*
Plessey PP250 processor *185*
Polynomial check codes *155*
Pre-selection *13, 113*
Presentation layer, OSI *162*
Privatisation *188*
Preventive cyclic re-transmission (#7) *159*
Protel UT 10/3 system *186, 195*
Protocol *6*
Pulse amplitude modulation (PAM) *12*
Pulse code modulation (PCM) *12*
Pulse generator *129*
Pulse VF signalling *138*
Push button signalling *139*
P wire *100, 105*

Quality, of transmission *23, 26*
Quantisation *32*
  noise *50, 52*
  non-linear *39*
  scoring *60*
  signal to Q noise ratio *50, 52*
  of μlaw *56*
Quasi-associated signalling *143*
  architecture *145*

Redundancy, elimination by coding *58*
Reed relays, cost *74*
Reeves, Alec *2, 12*
Reflection, of transmission *25*
Register *16, 104*
Register access *106*
Register control, functions *106*
Register signals *14, 16*
Release *13, 113*
Reliability, of exchanges *6*
Remote concentrator *94*
ROM *55*
Rotary system *104*
Routing *14, 17*
Routing signals *15, 16*
R2 signalling *139*

Sampling *29*
  spectrum *60*
Satellite transmission systems *22*
Scan control *106*
Scanning *14, 166*
SECAM *57*
Second chance, in pre-selection *88*
Segment, of coding law *54*
Selection *113*
Selection, permanent data *115*
Sequential read time switch *80*
Sequential write time switch *80*
Session layer, OSI *162*
Shannon, C. E. *2, 19*
Signalling *14, 134*
  AC *22*
  DC *22*

Signalling (contd.)
  in-band *22*
  out-band *22*
  systems *24*
Signal transfer point (STP) *144*
Simplex transmission *20*
Simple ended control, of synchronisation *133*
STP (signal transfer point) *144*
Slip *67, 131*
Slip rate *131*
Software
  control *112*
  data *113*
  modules *114*
  structure *119*
Space division (switching) *8*
Space switch stage *82*
SPC (stored program control) *17*
Speech memory *79, 92*
SSMF4 signalling *139*
STD (subscriber trunk dialling) *17*
Step-by-step switching *87*
  system (Strowger) *8, 100*
Stored program control (SPC) *17*
STP (signal transfer point) *144*
Strowger, A. B. *6*
Strowger step-by-step system *8, 100*
Structure, of software *119*
Submarine cable system *22*
Submerged repeater *22, 23*
Subscriber line circuit, step by step *100*
Subscriber line interface *90, 166*
Subscriber signalling, in PCM TDM *96*
Subscriber switching sub-system *90*
Subscriber trunk dialling (STD) *17*
Switch array architectures *84*
Switch control *106*
Switching architecture *86*

Switching
  blocking arrays *74*
  delay of TDM *85*
  non-blocking arrays *74*
  three-stage arrays *73*
Synchronisation *64, 66, 128, 130*
  despotic *66*
  mutual *66*
  signal, CEP channel 0 *135*
  strategies *66, 132*
  systems *132*
Synchronism *11*
System architecture *183*
System 1240 *112, 119, 178, 196*
System X *111, 116, 145, 185, 196*

TDM *1*
TDM switching *77*
Telephone user's pent (TUP) *151*
Temporary data *115*
Terminal switching *7*
Testing, of lines *93*
Thermionic valve *1, 6*
Third World, and the telephone *189*
Thirty-channel (CEPT) PCM system *44*
Thompson CSF MT range *110*
Three-stage switching arrays *72*
Time and space switch arrays *82*
Time division
  of control *99*
  of switching *7, 8, 11*
Time-slot *8*
Time-slot interchange *78*
Time switch *78*
  channel capacity *79*
  element *81*
  input associated *80*
  output associated *80*
  sequential read *80*

Time switch, sequential write *80*
Timing delay *64, 65*
Timing synchronisation *64*
Tone on idle signalling *137*
T1 PCM system *38, 39*
Traffic *4*
Transistor *2*
Transmission *20*
  duplex *20*
  simplex *20*
  switching *7*
  wire *20*
Transport layer, OSI *162*
Tropospheric scatter *22*
Trunk network interface *173*
TRW – Vidar *188*
TUP (telephone user's part) (#7) *151*
Twenty-four-channel PCM system
  D1 *38, 39*
  D2 *38, 40*
  UK *42*
Two motion selector *102*
TXE4A system *185*
TXE4 system *110, 129*
  switching network *76, 185*
TXE2 system *185*

UK CCS architecture *145*
UK 24-channel PCM system *42*
μ-law coding *42, 54*
μ-law coding table *55*
Uniform coding *50*
Uniselector *100*
US CCS architecture *145*
User signals *15*
UXD5 system *186, 197*

Variable length messages (#7) *151*
VF line signalling *137*
Virtual machine *119*